TOPICS IN MATHEMATICAL PHYSICS

PROBLEMY MATEMATICHESKOI FIZIKI

ПРОБЛЕМЫ МАТЕМАТИЧЕСКОЙ ФИЗИКИ

TOPICS IN MATHEMATICAL PHYSICS
Volume 1

SPECTRAL THEORY AND WAVE PROCESSES

Edited by
M. Sh. Birman
Department of Physics
Leningrad State University

Translated from Russian

c/b CONSULTANTS BUREAU · NEW YORK · 1967

First Printing – March 1967
Second Printing – July 1969

ISBN 978-1-4684-7597-5 ISBN 978-1-4684-7595-1 (eBook)
DOI 10.1007/978-1-4684-7595-1

The original Russian text was published by Leningrad University Press in 1966.

Проблемы математической физики
Выпуск 1
Спектральная теория и волновые процессы

Library of Congress Catalog Card Number 67-16365

PREFACE

The articles in this collection are devoted to various problems in mathematical physics and mathematical analysis, primarily in the fields of spectral theory and the theory of wave processes.

The collection is intended for mathematicians specializing in the fields of mathematical physics, functional analysis, and the theory of differential equations. In addition, it is of some interest to theoretical physicists.

The first paper deals with a mixed boundary value problem for a system of elasticity equations, and considers fields in the neighborhood of various wave fronts. The method used permits an estimate of the errors in the Ben-Menahem approximate method.

Paper 2 investigates operators in separable Hilbert space given by double integrals of a type defined at the beginning of the paper, and in which integration is understood as the limit of the integral sums of Riemann-Stieltjes.

In paper 3, the problem of calculation of elastic constants for a laminarly inhomogeneous semi-infinite medium is considered, and the uniqueness of the solution of the inverse seismic problem at finite depth proved.

The fourth paper gives a detailed account of the results of an earlier paper by the same author, in which he generalized to the three-dimensional case the trace formulas obtained for the one-dimensional Schroedinger operator. Asymptotic estimates of the resolvent kernel and solutions of the scattering problem are given.

The last paper considers nonself-adjoint differential operators – in particular, the one- and three-dimensional Schroedinger operators. A method for overcoming the difficulty in the expansion in eigenfunctions of these operators is proposed.

CONTENTS

THE LAMB PROBLEM FOR AN INHOMOGENEOUS ELASTIC HALF-SPACE

A. G. Alenitsyn

§ 1. Formulation of the Problem

The following mixed boundary value problem for the system of equations of elasticity is considered: To find the displacement vector $\vec{u}(x, y, z, t) = (u_x, u_y, u_z)$ satisfying in $-\infty < x < +\infty$, $-\infty < y < +\infty$, $z > 0$, $t > 0$ the equations of dynamics of an isotropic elastic medium

$$\rho\vec{u}_{tt} = (\lambda + 2\mu)\nabla(\nabla, \vec{u}) - \mu[\nabla, [\nabla, \vec{u}]] + \nabla\lambda(\nabla, \vec{u}) + 2(\nabla\mu, \nabla)\vec{u} + [\nabla\mu, [\nabla, \vec{u}]], \tag{1.1}$$

where the boundary conditions at $z = 0$ are

$$\tau_{zx} \equiv \mu\left(\frac{\partial u_x}{\partial z} + \frac{\partial u_z}{\partial x}\right) = 0, \quad \tau_{zy} \equiv \mu\left(\frac{\partial u_y}{\partial z} + \frac{\partial u_z}{\partial y}\right) = 0,$$
$$\tau_{zz} \equiv 2\mu\frac{\partial u_z}{\partial z} + \lambda(\nabla, \vec{u}) = \delta(x)\delta(y)\varepsilon(t), \tag{1.2}$$

and the initial conditions at $t = 0$ are

$$\vec{u} = \vec{u}_t = 0. \tag{1.3}$$

Here λ and μ are the Lamé coefficients; ρ is the density; τ_{zx}, τ_{zy}, τ_{zz} are the components of the stress tensor; $\varepsilon(t)$ is the Heaviside function; and $\delta(x)$ is the Dirac function.

The system (1.1)-(1.3) will be called the Lamb problem for the system of equations of elasticity.

The two-dimensional problem will be called a planar Lamb problem when the system (1.1)-(1.3) is subject to the additional conditions: $u_y \equiv 0$, the displacement vector $\vec{u}$ is independent of y, and formulas (1.2) are replaced by the formulas

$$\tau_{zx} = 0, \quad \tau_{zz} = \delta(x)\varepsilon(t). \tag{1.4}$$

Let us introduce the cylindrical coordinates r, φ, z. The problem (1.1)-(1.3) will be called an axially symmetric Lamb problem when $u_\varphi \equiv 0$, the vector $\vec{u}$ is independent of φ, and formulas (1.2) are replaced by the formulas

$$\tau_{zr} \equiv \mu\left(\frac{\partial u_r}{\partial z} + \frac{\partial u_z}{\partial r}\right) = 0, \quad \tau_{zz} = \frac{1}{r}\delta(r)\varepsilon(t). \tag{1.5}$$

Problem (1.1)-(1.3), formulated in a somewhat different form by Lamb [1], has been considered in a large number of papers. The first rigorous solution of the planar and axially symmetric Lamb problems was given by Academicians Smirnov and Sobolev [2, 3]. Unfortunately, the Smirnov–Sobolev method (the method of functionally invariant solutions) is not applicable to inhomogeneous media even in the particular, but very important, practical case when the coefficients of the equations depend only on a single coordinate.

Petrashen' [4] has solved the planar Lamb problem for a uniform elastic half space with the help of a separation of variables. A detailed investigation of the solutions of the planar and axially symmetric Lamb problems for a homogeneous half space has been made in [5]. The method of separation of variables allows us to obtain and investigate the solution of the Lamb problem also in the case when the coefficients of the equations depend on the z coordinate. The axially symmetric problem of the type of the Lamb problem for a scalar wave equation has been considered by Alekseev [6]; here, the equation to be solved is

$$u_{tt}=v^2(z)\Delta u, \tag{1.6}$$

where the velocity v(z) is a linearly increasing function. The case of a monotonically decreasing analytic function v(z) has been investigated by Molotkov [7]; Molotkov and Mukhina [8] have also investigated the case when v(z) has a single minimum, but is otherwise monotonic.

Various approximate methods, in particular, the ray method, can be used in the general case of an inhomogeneous medium for the solution of the Lamb problem. The ray method can be used to calculate the displacement field in the illuminated part of the half space [9] and the field of the Rayleigh waves [10], but it is not possible to calculate the field in the shadow region and the field of a shadow-type head wave.

In the present paper, the method of separation of variables is used to derive the exact solution (§ 2) of the planar and axially symmetric Lamb problems for the system of equations of elasticity (1.1) when the coefficients λ, μ, ρ are functions of a single coordinate z. The investigation of the exact solution in the neighborhood of the fronts of various waves requires the determination of the asymptotic solutions of an auxiliary system of ordinary differential equations containing a large parameter. This aspect of the problem is considered in § 3, in which this asymptotic behavior is derived with the help of a simple modification of the classical Birkhoff–Tamarkin asymptotic theory [11] and the Sibuya theorem [12] on asymptotic expansions. Starting with § 4, it is everywhere assumed in addition that the velocities $v_p(z)$ of the longitudinal and $v_s(z)$ of the transverse waves [$v_p^2 = (\lambda + 2\mu)/\rho$, $v_s^2 = \mu/\rho$] are monotonically decreasing functions. In §§ 5, 7, 8, and 9 the coefficients $\lambda(z)$, $\mu(z)$, and $\rho(z)$ are assumed to be analytic for $z \geq 0$.* The equations of the wave fronts of the various waves are given in § 4. The exact solution is then used for the calculation of the first term in the asymptotic expansion of the nonanalytic part of the field in the vicinity of the wave fronts of the longitudinal and transverse waves in the illuminated part of the half space (§ 5) and in the vicinity of the Rayleigh wave front (§ 6); the asymptotic formulas for the fields of the longitudinal and transverse waves in the shadow region are obtained in § 7, while in § 8 the analogous calculation is carried out for the field of the head wave; § 9 contains the corresponding formulas for the axially symmetric Lamb problem, and in § 10 some critical remarks are made concerning the approximate Ben-Menahem method [13].

*In addition to these restrictions, some additional restrictions on the behavior of λ, μ, ρ at infinity are imposed in this paper.

§ 2. Exact Solution

Let us consider the planar Lamb problem for the system (1.1). We will look for a solution $\vec{u}(x, z, t) = \begin{pmatrix} u_1 \\ u_2 \end{pmatrix}^*$ in the form

$$\vec{u}(x, z, t) = \frac{1}{\pi}\int_0^\infty dk \frac{1}{2\pi i}\int_M \begin{pmatrix} G_1(z, k, s)\sin kx \\ G_2(z, k, s)\cos kx \end{pmatrix} \frac{e^{kts}}{s}\, ds, \tag{2.1}$$

where $\begin{pmatrix} G_1(z, k, s) \\ G_2(z, k, s) \end{pmatrix} = \vec{G}(z, k, s)$ is an unknown vector and M is the Mellin contour.

The substitution of (2.1) into (1.1) yields a system of two ordinary differential equations for the vector $\vec{G}$

$$\vec{G}'' + k\begin{Vmatrix} 0 & 1-p^{-1} \\ 1-p & 0 \end{Vmatrix}\vec{G}' - k^2\begin{Vmatrix} p^{-1}m_p^2 & 0 \\ 0 & pm_s^2 \end{Vmatrix}\vec{G} + \begin{Vmatrix} \mu'/\mu & 0 \\ 0 & \nu'/\nu \end{Vmatrix}\vec{G}' + k\begin{Vmatrix} 0 & -\mu'/\mu \\ \lambda'/\lambda & 0 \end{Vmatrix}\vec{G} = 0. \tag{2.2}$$

Here, $\nu = \nu(z) \equiv \lambda(z) + 2\mu(z)$, $p = p(z) \equiv \mu(z)/\nu(z) < 1/2$, $m_n^2(z, s) \equiv 1 + n_p^2(z)s^2$, $m_s^2(z, s) \equiv 1 + n_s^2(z)s^2$, $n_p^2(z) \equiv \rho(z)/\nu(z)$, $n_s^2(z) \equiv \rho(z)/\mu(z)$, and a prime denotes differentiation with respect to z.

It is easy to see that the boundary conditions (1.4) are satisfied if

$$\vec{G} = \left(D_s\vec{G}^{(p)} - D_p\vec{G}^{(s)}\right)\nu_0^{-1}\Delta^{-1}, \tag{2.3}$$

where

$$\Delta(k, s) = E_pD_s - E_sD_p, \quad \nu_0 = \nu(0),$$
$$D_p(k, s) = \left(G_1^{(p)'} - kG_2^{(p)}\right)\Big|_{z=0}, \quad D_s(k, s) = \left(G_1^{(s)'} - kG_2^{(s)}\right)\Big|_{z=0},$$
$$E_p(k, s) = \left(G_2^{(p)'} + k\frac{\lambda}{\nu}G_1^{(p)}\right)\Big|_{z=0}, \quad E_s(k, s) = \left(G_2^{(s)'} + k\frac{\lambda}{\nu}G_1^{(s)}\right)\Big|_{z=0}$$

the vectors $\vec{G}^{(p)}$ and $\vec{G}^{(s)}$ are any two linearly independent solutions of system (2.2). The choice of these vectors will be made in § 3 in such a manner that the initial conditions (1.3) are satisfied.

Let us now consider the axially symmetric Lamb problem. Formula (2.1) in this case is replaced by

$$\vec{u}(r, z, t) = \int_0^\infty kdk \frac{1}{2\pi i}\int_M \begin{pmatrix} G_1(z, k, s)J_1(kr) \\ G_2(z, k, s)J_0(kr) \end{pmatrix} \frac{e^{kts}}{s}\, ds, \tag{2.4}$$

where J_0 and J_1 are Bessel functions, while the vector $\vec{G}(z, k, s)$ is the same as that in the planar problem.

§ 3. The Asymptotic Behavior of the Solutions of System (2.2)

In order to investigate the displacement field $\vec{u}$, it is necessary to investigate the asymptotic behavior of the solution $\vec{G}$ of system (2.2) as $k \to \infty$.

---

* Here and in the following the symbol $\begin{pmatrix} f_1 \\ f_2 \end{pmatrix}$ denotes the vector with the components f_1 and f_2.

It is convenient to write (2.2) as a system

$$\vec{Z}' = [kH(z, s) + K(z)]\vec{Z} \tag{3.1}$$

of four differential equations in which

$$\vec{Z} = \begin{pmatrix} G_1 \\ G_2 \\ k^{-1}G_1' \\ k^{-1}G_2' \end{pmatrix}. \tag{3.2}$$

The matrices H and K are

$$H = \begin{Vmatrix} 0 & 0 & 1 & 0 \\ 0 & 0 & 0 & 1 \\ \frac{m_p^2}{p} & 0 & 0 & \frac{1}{p} - 1 \\ 0 & pm_s^2 & p-1 & 0 \end{Vmatrix}, \quad K = \begin{Vmatrix} 0 & 0 & 0 & 0 \\ 0 & 0 & 0 & 0 \\ 0 & \frac{\mu'}{\mu} & -\frac{\mu'}{\mu} & 0 \\ -\frac{\lambda'}{\nu} & 0 & 0 & -\frac{\nu'}{\nu} \end{Vmatrix}$$

In the following, we will frequently encounter the eigenvalues $\omega_j(z, s)$ $(j = 1, 2, 3, 4)$ of the matrix $H(z, s)$, which are equal to $\mp m_p(z, s)$ and $\mp m_s(z, s)$; we will assume that the plane s has been cut along the imaginary axis between the points $\pm iv_p(z)$ and that the branches of the functions $m_p(z, s)$ and $m_s(z, s)$ are fixed by the conditions $\arg m_{p,s} = 0$ and $\arg s = 0$.

The classical asymptotic theory of Birkhoff–Tamarkin for linear systems of ordinary differential equations containing a large parameter is well established (for example, see [14]). However, the simple formulas given by the classical theory are valid only under the following restrictions which are formulated here in connection with system (3.1):

1) $z \in [a, b]$, where a and b are finite;

2) $\omega_i \neq \omega_j$, when $i \neq j$, over the whole interval $[a, b]$;

3) the quantities $\mathrm{Re}(\omega_i - \omega_j)$, with i and j fixed, do not change their sign in $[a, b]$;

4) the matrices H and K are sufficiently smooth functions of z;

5) the parameter s is fixed.

Condition (5) can be relaxed by letting the parameter s vary inside a finite closed region S, at all points of which conditions (2) and (3) are satisfied. This theory is effective for those problems where it is sufficient to know the asymptotic behavior of the solution as $k \to \infty$ when the parameter s varies in the neighborhood of some point in the complex plane s. An example of such a problem is the problem of stationary Rayleigh waves in a half space [15]. However, in our nonstationary problem we require the asymptotic behavior of the solution $\vec{Z}$ essentially over the whole complex plane s. We will have to modify the classical procedure for the derivation of the asymptotic representation of the solution of system (3.1) in order to extend the classical theory to wider regions than those that are allowed by conditions (1) and (3). Moreover, we will consider separately the neighborhood of those points (reversal points) where $\omega_i = \omega_j$, i.e., condition (2) is violated.

Let $Z(z, k, s)$ be a matrix solution of system (3.1). We make the change of variable $Z = UX$, where $U(z, s)$ is a nonsingular matrix, and obtain

$$X' = (k\widetilde{H} + \widetilde{K})X, \quad \widetilde{H} = U^{-1}HU, \quad \widetilde{K} = U^{-1}KU - U^{-1}U. \tag{3.3}$$

Let us exclude from consideration for the time being the neighborhoods of the points $s = 0$ and $s = \pm i v_p(z)$, $\pm i v_s(z)$ for all $z \geq 0$, so that condition (2) is satisfied for all z. Let $U^{-1}HU = \Lambda = [-\mathbf{m}_p, \mathbf{m}_p, -\mathbf{m}_s, \mathbf{m}_s]$ be the diagonal form of the matrix H. We take matrix U to be

$$U = \left\| \begin{matrix} 1 & 1 & m_s & m_s \\ m_p & -m_p & 1 & -1 \\ -m_p & m_p & -m_s^2 & m_s^2 \\ -m_p^2 & -m_p^2 & -m_s & -m_s \end{matrix} \right\|$$

As is known [14], when condition (2) is satisfied there exists a formal solution of system (3.3) of the form

$$\sum_{\nu=0}^{\infty} T_\nu(z, s) k^{-\nu} \exp\left\{\int_0^z (k\Lambda(\zeta, s) + Q(\zeta, s))\, d\zeta\right\}. \tag{3.4}$$

The matrices $T_\nu(z, s)$ and the diagonal matrix $Q(z, s)$ are determined by recurrence formulas and are found to be infinitely differentiable, provided that H and K are infinitely differentiable. In particular,

$$T_0 = E^*, \quad T_1\Lambda - \Lambda T_1 = \widetilde{K} - Q, \tag{3.5}$$

so that we have

$$(T_1)_{ij} = \frac{\widetilde{K}_{ij}}{\omega_j - \omega_i} \quad (i \neq j), \quad Q_{ii} = \widetilde{K}_{ii}. \tag{3.6}$$

As is usual, let us write down the expression

$$\hat{X}(z, k, s) \equiv \sum_{\nu=0}^{m} T_\nu(z, s) k^{-\nu} \exp\left\{\int_0^z (k\Lambda + Q)\, d\zeta\right\}. \tag{3.7}$$

This expression satisfies the system of equations

$$\hat{X}' = \left(k\Lambda + \widetilde{K} + k^{-m}M\right)\hat{X}, \tag{3.8}$$

where the matrix $M = M(z, k, s)$ can be expressed in terms of $T_1, \ldots, T_m$ and their derivatives and is bounded as $k \to \infty$ in any finite interval $z \in [a, b]$. Since we will require to know the uniform asymptotic behavior over the semi-axis $0 \leq z \leq \infty$ of the solution of system (3.1), let us evaluate the matrix M as $z \to \infty$. Here, we will carry out the derivation for the case $m = 1$ as this is sufficient for most practical purposes; evaluations for $m > 1$ are obtained in an analogous manner.

The expression for the matrix M is

$$M = \left(T_1' + T_1 Q - \widetilde{K} T_1\right)(E + k^{-1} T_1)^{-1}. \tag{3.9}$$

We will assume that the functions $\lambda(z)$, $\mu(z)$, and $\rho(z)$ have two continuous derivatives for $z \geq 0$ and that the following conditions are satisfied:

$$\lambda'', \mu'', \rho'' = O(z^{-1-\alpha_1}), \quad z \to \infty, \quad \alpha_1 > 0, \tag{3.10}$$

*E is the unit matrix.

$$\lambda', \mu', \rho' = O\left(z^{-\frac{1}{2}-\alpha_2}\right), \quad z \to \infty, \quad \alpha_2 > 0, \tag{3.10}$$

$$0 < \alpha_3 \leqslant \lambda, \mu, \rho \leqslant \alpha_4 < \infty. \tag{3.11}$$

From conditions (3.10) and (3.11), we see that U and U^{-1} are O(1); U', K, $\widetilde{K}$, Q, T_1 are $O(z^{(-1/2)(-\alpha_2)})$; and T_1' is $O(z^{-1-\alpha_1})$, so that, at least for sufficiently large k, we have

$$M = O(z^{-1-\varepsilon}), \quad z \to \infty, \quad \varepsilon > 0. \tag{3.12}$$

The evaluation of the matrix $\mathcal{E} \equiv \exp\left\{\int_0^z Q(\zeta, s)\, d\zeta\right\}$ yields the formula

$$\mathcal{E}^2 = \frac{\rho(0)}{\rho(z)}\left[\frac{m_p(0, s)}{m_p(z, s)}, \frac{m_o(0, s)}{m_p(z, s)}, \frac{m_s(0, s)}{m_s(z, s)}, \frac{m_s(0, s)}{m_s(z, s)}\right], \tag{3.13}$$

which shows that $\mathcal{E}$ and $\mathcal{E}^{-1}$ are bounded.

We make the change of variable $X = (E + k^{-1}T_1)\mathcal{E}F$. An integral equation can be easily derived for the matrix F(z, k, s) which can be written as

$$\begin{aligned} F(z, k, s) = {} & \exp\left(k\int_0^z \Lambda\, d\zeta\right) - k^{-1}\int_0^z K_1(z, z', k, s)\, F(z', k, s)\, dz' \\ & - k^{-1}\int_\infty^z K_2(z, z', k, s)\, F(z', k, s)\, dz', \end{aligned} \tag{3.14}$$

where

$$K_1(z, z', k, s) = \left[e^{-k\int_{z'}^z m_p d\zeta},\ 0,\ e^{-k\int_{z'}^z m_s d\zeta}\ 0\right]\widetilde{M}(z', k, s),$$

$$K_2(z, z', k, s) = \left[0,\ e^{k\int_{z'}^z m_p d\zeta}\ 0,\ e^{k\int_{z'}^z m_s d\zeta}\right]\widetilde{M}(z', k, s),$$

$$\widetilde{M} = \mathcal{E}^{-1}(E + k^{-1}T_1)^{-1}M(E + k^{-1}T_1)\,\mathcal{E}.$$

If $\operatorname{Re} s \geq \sigma_0 > 0$, then $\operatorname{Re} m_{p,s} \geq \sigma_1 > 0$ with our choice of branches, and as $k \to +\infty$ the integral operator on the right-hand side of (3.14) in the space $C[0, \infty)$ has a norm tending to zero. Of the four vectors composing the free term of Eq. (3.14), the first and third are bounded for $\operatorname{Re} s \geq 0$, so that we obtain the asymptotic formula

$$\vec{F}^{(j)}(z, k, s) = \vec{e}_j \exp\left\{k\int_0^z \omega_j(\zeta, s)\, d\zeta\right\} + k^{-1}O(1) \quad (j = 1, 3), \tag{3.15}$$

which is correct for $\operatorname{Re} s \geq \sigma_0 > 0$, $|s| < \infty$, $k \to +\infty$, $z \in [0, \infty)$. Here, $\vec{e}_j$ is the j-th unit vector in four-dimensional space.

It is important to note that for those s for which Λ is real, the solutions of Eq. (3.14) and system (3.1) obtained by the method described above are real. Since $\operatorname{Re} m_{p,s} > 0$ also on the right-hand edge of the segment $L = [i\varepsilon, iv_i - i\varepsilon]$, where $v_i \equiv \inf\limits_{z \in [0, \infty)} v_s(z)$ and $\varepsilon > 0$, then formula (3.15) is correct right up to the right-hand edge of L (and the segment conjugate to it). The vectors $\vec{F}^{(1)}$ and $\vec{F}^{(3)}$ are real on this segment which will be found to be important in § 5.

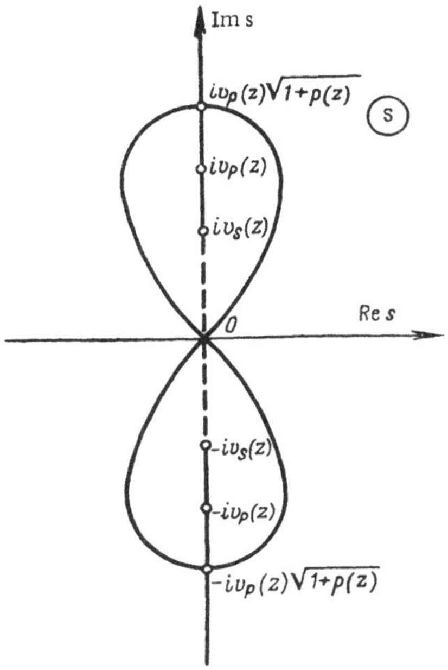

Fig. 1

In a certain region of the right-hand half-plane it is possible to derive the usual classical asymptotic expression from (3.14) and to do this, it is sufficient to multiply this equation by $\exp\left(-k\int_0^z \omega_j d\zeta\right)$ and to introduce appropriate new unknown vectors. Let us consider this in greater detail.

The lines $\mathrm{Re}\,(\omega_i - \omega_j)$ which are critical lines in the Birkhoff–Tamarkin theory, are the following in our case:

1) the lines $\mathrm{Re}\, m_{p,s}(z, s) = 0$ are parts of the imaginary axis leading to $\pm i\infty$ from the points $\pm iv_{p,s}(z)$, respectively;

2) the lines $\mathrm{Re}\, m_{-}(z, s) = 0$ ($m_{\pm} \equiv m_p \pm m_s$) form a figure-eight curve with center at the origin; they leave the origin at an angle of $\pi/4$ to the imaginary axis and intersect it at right angles (Fig. 1);

3) the lines $\mathrm{Re}\, m_{+}(z, s) = 0$ coincide with the lines $\mathrm{Re}\, m_p(z, s) = 0$.

Segments of the imaginary axis between the points $\pm iv_s(z)$ are also critical lines, inasmuch as these segments lie on the cut.

Let us use $S_1(z)$ to denote the part of the half-space $\mathrm{Re}\, s > 0$ which is to the right of the line $\mathrm{Re}\, m_{-}(z, s) = 0$ and to the right of the circle $|s| = \varepsilon > 0$. In view of (3.11), $0 < c_1 \leq v_{p,s} \leq c_2 < \infty$; therefore, there exists a section S_1 common to all of these regions $S_1(z)$ for all $z \geq 0$.

If $s \in S_1$, then $\mathrm{Re}\, m_{-}(z, s) < 0$ for all $z \geq 0$; therefore, as $k \to +\infty$ the integral equation (3.14) yields

$$\vec{F}^{(1)}(z, k, s) = \left(\vec{e}_1 + k^{-1}O(1)\right) \exp\left\{-k\int_0^z m_p(\zeta, s)\, d\zeta\right\}. \tag{3.16}$$

This formula is valid when $s \in S_1$ and is uniform over z for $z \in [0, \infty)$.

Similarly, let $S_3(z)$ be the part of the right-hand half-space included between the line $\mathrm{Re}\, m_{-} = 0$ and the imaginary axis with the exception of the regions

$$|s \pm iv_p(z)| \leqslant \varepsilon, \quad |s \pm iv_s(z)| \leqslant \varepsilon, \quad |s| \leqslant \varepsilon. \tag{3.17}$$

In the common section S_3 of all regions $S_3(z)$, the following asymptotic formula holds for $z \geq 0$:

$$\vec{F}^{(3)}(z, k, s) = \left(\vec{e}_3 + k^{-1}O(1)\right) \exp\left\{-k\int_0^z m_s(\zeta, s)\, d\zeta\right\}. \tag{3.18}$$

Formulas of type (3.15), (3.16), and (3.18) will be called classically asymptotic formulas. It should be noted that with an appropriate strengthening of condition (3.10) we would have obtained for $m > 1$ a correction of the form $k^{-m}O(1)$ in these formulas.

Let us now consider the neighborhoods of the reversal points. In order to obtain the asymptotic solutions of system (3.1) we can make use of the general theory developed by Sibuya [12].

Let us consider the system

$$\varepsilon \vec{y}'(t) = A(t, \varepsilon)\vec{y}(t) \tag{3.19}$$

of n equations with square matrix $A(t, \varepsilon)$ regular in the region $t \in D$, $\varepsilon \in \mathcal{E}$, where D is the neighborhood $|t - t_0| < \delta_0$ of a fixed point t_0, the region $\mathcal{E}$ is defined by the inequality $0 < |\varepsilon| \leq \varepsilon_0$, $|\arg \varepsilon| \leq \alpha_0$. It is assumed that $A(t, \varepsilon)$ can be expanded for $\varepsilon \to 0$ into the asymptotic series

$$A(t, \varepsilon) \sim \sum_{\nu=0}^{\infty} \varepsilon^\nu A_\nu(t) \tag{3.20}$$

with matrices $A_\nu(t)$ regular in D. We can assume that the matrix $A_0(t_0)$ has been reduced to the Jordan canonical form.

Sibuya's Theorem

In the subdomain $D_1 \subset D$, $\mathcal{E}_1 \subset \mathcal{E}$ defined by the inequalities

$$|t - t_0| < \delta_1 \leqslant \delta_0, \quad 0 < |\varepsilon| \leqslant \varepsilon_1 \leqslant \varepsilon_0, \quad |\arg \varepsilon| \leqslant \alpha_1 \leqslant \alpha_0, \tag{3.21}$$

there exists a regular nonsingular transformation $\vec{y}(t, \varepsilon) = P(t, \varepsilon)\vec{\Omega}(t, \varepsilon)$, such that in an appropriate system for $\vec{\Omega}$,

$$\varepsilon \vec{\Omega}' = B(t, \varepsilon)\vec{\Omega} \quad (B = P^{-1}AP - \varepsilon P^{-1}P'); \tag{3.22}$$

the matrix $B(t, \varepsilon)$ is quasidiagonal: $B(t, \varepsilon) = [\{B(t, \varepsilon)\}_1, \ldots, \{B(t, \varepsilon)\}_s$, $(s \leq n)$; this matrix has the same structure as the matrix $A_0(t_0)$. $P(t, \varepsilon)$ and $B(t, \varepsilon)$ can be expanded into the asymptotic series

$$P(t, \varepsilon) \sim P_0(t) \sum_{\nu=0}^{\infty} \varepsilon^\nu S_\nu(t), \qquad B(t, \varepsilon) \sim \sum_{\nu=0}^{\infty} \varepsilon^\nu B_\nu(t), \tag{3.23}$$

where the matrices $P_0(t)$, $S_\nu(t)$, and $B_\nu(t)$ are regular in D_1 and can be calculated recurrently. In particular, $P_0(t)$ transforms $A_0(t)$ into quasidiagonal form:

$$P_0^{-1}(t)A_0(t)P_0(t) = B_0(t) = [(B_0(t))_1, \ldots, (B_0(t))_s], \; P_0(t_0) = E, \; S_0(t) = E. \tag{3.24}$$

We will require that the functions $\lambda(z)$, $\mu(z)$, and $\rho(z)$ be regular in an interval $z \in (-\beta, b)$ $(\beta, b > 0)$ and we will apply Sibuya's theorem, assuming that $\varepsilon = k^{-1}$, $t = z$, $A_0(t) = H(z, s)$ (s fixed), $A_1(t) = K(z)$, $A_2 = A_3 = \ldots = 0$. It can be shown that the theorem holds also when s varies in some neighborhood of any point $s \neq 0$.

The reversal points for system (3.1) are determined by the equations $\pm i v_{p,s}(z) = s$. We will assume that the velocities $v_p(z)$ of the longitudinal and $v_s(z)$ of the transverse waves are strictly monotonic for $z \in (-\beta, b)$, and that $v_p' \neq 0$ and $v_s' \neq 0$. The regions $|s - iv_p(z)| \leq \varepsilon$ and $|s - iv_s(z)| \leq \varepsilon$ of the s plane we will denote by $F_p(z)$ and $F_s(z)$, respectively; $F_p^*(z)$ and $F_s^*(z)$ are the regions symmetric to these in the lower half-plane.

Let us consider region $F_p(z)$. A simple calculation shows that in this region

$$P_0(z, s) = \begin{Vmatrix} 1 & 0 & m_s & m_s \\ 0 & -1 & 1 & -1 \\ 0 & 1 & -m_s^2 & m_s^2 \\ -m_p^2 & 0 & -m_s & -m_s \end{Vmatrix}, \tag{3.25}$$

$$B_0(z, s) = \begin{Vmatrix} 0 & 1 & 0 & 0 \\ m_p^2 & 0 & 0 & 0 \\ 0 & 0 & -m_s & 0 \\ 0 & 0 & 0 & m_s \end{Vmatrix} \tag{3.25}$$

For the first block Ω_1 of the fundamental matrix Ω of system (3.22) we have a system of equations

$$\Omega_1' = k(B(z, k, s))_1\Omega_1, \tag{3.26}$$

where

$$(B)_1 \sim \begin{Vmatrix} 0 & 1 \\ m_p^2 & 0 \end{Vmatrix} + k^{-1}\begin{Vmatrix} a & 0 \\ 0 & b \end{Vmatrix} + \ldots,$$

$$a(z, s) = -\ln'\rho - \frac{2}{n_s^2 s^2}\ln'\mu, \quad b(z, s) = \frac{2}{n_s^2 s^2}\ln'\mu, \quad \ln' f = \frac{f'}{f}.$$

For the second and third blocks of matrix Ω we have the equations

$$\Omega_j' = k(B(z, k, s))_j\Omega_j \quad (j = 2, 3), \tag{3.27}$$

where

$$(B)_j \sim (-1)^{j-1}m_s + k^{-1}c + \ldots,$$
$$c(z, s) = -\frac{1}{2}\ln'\rho m_s.$$

The solutions of Eqs. (3.27) are obviously the following:

$$\Omega_j = \sqrt{\frac{\rho(0)\, m_s(0, s)}{\rho(z)\, m_s(z, s)}}\left[\sum_{\nu=0}^{m-1} k^{-\nu} f_{\nu j}(z, s) + O(k^{-m})\right]\exp\left\{(-1)^{j-1}k\int_0^z m_s d\zeta\right\}. \tag{3.28}$$

Here $f_{0j} = 1$, $f_{1j}, \ldots, f_{m-1,j}$ are determined by recurrence formulas, and $j = 2, 3$.

System (3.26) can be reduced to a second-order equation and use made of Olver's results [16] or, alternatively, we can obtain the asymptotic solution of system (3.26) with the help of the method of the standard equation. We obtain the following asymptotic formula for the fundamental matrix Ω_1 of system (3.26):

$$\Omega_1(z, k, s) = R_0(z, s)\left[\sum_{\nu=0}^{m-1} V_\nu(z, s)k^{-\nu} + O(k^{-m})\right]W(z, k, s), \tag{3.29}$$

where

$$R_0(z, s) = E\exp\left\{\frac{1}{2}\int_0^z (a+b)\,d\zeta\right\} = \sqrt{\frac{\rho(0)}{\rho(z)}}E, \quad V_0 = E;$$

$V_1, \ldots, V_{m-1}$ are given by the recurrence formulas

$$W(z, k, s) = \begin{Vmatrix} y_1 & y_2 \\ k^{-1}y_1' & k^{-1}y_2' \end{Vmatrix}, \quad y_j = \frac{1}{\sqrt{\varphi'(z, s)}}Ai_j\left(k^{2/3}\varphi(z, s)\right) \; (j = 1, 2),$$

$$\varphi(z,\ s)=\left(\frac{3}{2}\int\limits_{z_p(s)}^{z} m_p(\zeta,\ s)\,d\zeta\right)^{2/3}.$$

Ai_1 and Ai_2 are any two linearly independent solutions of the Airy equation $(d^2/dt^2)Ai(t) = tAi(t)$, the prime denotes differentiation with respect to z, and $z_p(s)$ is the reversal point satisfying the equation $m_p^2(z, s) = 0$.

Formulas (3.28) and (3.29) are valid at least in the domain $F_p(z)$. Thus, the fundamental matrix Z of system (3.1) in $F_p(z)$ is given by

$$Z(z,\ k,\ s)=P_0(z,\ s)\left[\sum_{\nu=0}^{m-1} k^{-\nu}M_\nu(z,\ s)+O(k^{-m})\right]Y(z,\ k,\ s), \tag{3.30}$$

where

$$Y=\left[W,\ \sqrt{\frac{m_{s0}}{m_s}}\,e^{-k\int\limits_0^z m_s d\zeta},\ \sqrt{\frac{m_{s0}}{m_s}}\,e^{k\int\limits_0^z m_s d\zeta}\right],\quad m_{s0}\equiv m_s(0,\ s).$$

$$M_0=\sqrt{\frac{\rho(0)}{\rho(z)}}\,E;$$

$M_1, \ldots, M_{m-1}$ are calculated from formulas (3.24)-(3.29) (here, it will also be sufficient for us to take $m = 1$).

Formulas analogous to (3.30) hold in domain $F_s(z)$, as well as in the conjugate domain.

We take the solutions $\vec{G}^{(p)}$ and $\vec{G}^{(s)}$ (see § 2) to be vectors composed of the two first components of the vectors $\vec{Z}^{(1)}$ and $\vec{Z}^{(3)}$, where

$$\vec{Z}^{(j)}(z,\ k,\ s)=U(z,\ s)(E+k^{-1}T_1(z,\ s))\,\mathcal{E}(z,\ s)\vec{F}^{(j)}(z,\ k,\ s)\ (j=1,3). \tag{3.31}$$

It is not difficult to show that in fact the remainder terms in formulas (3.15) and (3.16) are indeed of the form $(ks)^{-1}O(1)$ and are uniform over s for $\mathrm{Re}\, s \geq \sigma_0 > 0$, from which it follows that for $t < 0$ the exact solution is identically zero and the initial conditions are satisfied.

In addition to the solutions $\vec{Z}^{(1)}$ and $\vec{Z}^{(3)}$, we will now examine other solutions. Let $\vec{Z}_*^{(1)} = U(E + k^{-1}T_1)\,\mathcal{E}\vec{F}_*^{(1)}$, where the vector $\vec{F}_*^{(1)}(z, k, s)$ is given by the integral equation

$$F_*^{(1)}e^{k\int\limits_0^z m_p d\zeta}=\vec{e}_1-k^{-1}\int\limits_\infty^z\left[1,\ e^{2k\int\limits_{z'}^{z} m_p d\zeta},\ e^{k\int\limits_{z'}^{z} m_- d\zeta},\ e^{k\int\limits_{z'}^{z} m_+ d\zeta}\right]\widetilde{M}\vec{F}_*^{(1)}e^{k\int\limits_0^{z'} m_p d\zeta}\,dz'. \tag{3.32}$$

This equation defines the solution $\vec{F}_*^{(1)}$ in domain S_3 and gives for it the following asymptotic formula:

$$\vec{F}_*^{(1)}(z,\ k,\ s)=\left(\vec{e}_1+k^{-1}O(1)\right)e^{-k\int\limits_0^z m_p d\zeta}. \tag{3.33}$$

The solution $\vec{F}_*^{(1)}$ is real on segment L.

Similarly, in domain S_1 we define vector $\vec{F}_*^{(3)}$ by the integral equation

$$\vec{F}_*^{(3)} e^{k\int_0^z m_s d\zeta} = \vec{e}_3 - k^{-1} \int_\infty^z \left[e^{-k\int_{z'}^z m_- d\zeta},\ e^{k\int_{z'}^z m_+ d\zeta}\ 1,\ e^{2k\int_{z'}^z m_s d\zeta} \right] \widetilde{M} \vec{F}_*^{(3)} e^{k\int_0^{z'} m_s d\zeta} dz', \tag{3.34}$$

giving in S_1 the formula

$$\vec{F}_*^{(3)}(z,\ k,\ s) = \left(\vec{e}_3 + k^{-1} O(1)\right) e^{-k\int_0^z m_s d\zeta}. \tag{3.35}$$

This solution is real for $s > 0$.

The following lemma can be easily proved.

Lemma

$$\vec{Z}^{(1)}(z,\ k,\ s) \equiv \alpha(k,\ s)\, \vec{Z}_*^{(1)}(z,\ k,\ s) + \beta(k,\ s)\, \vec{Z}^{(3)}(z,\ k,\ s), \tag{3.36}$$

where α and β are functions regular in the half-plane $\operatorname{Re} s > 0$ for fixed $k \gg 1$.

In the left-hand half-plane, the vectors bounded as $z \to \infty$ are $\vec{Z}^{(2)}$ and $\vec{Z}^{(4)}$ defined analogously to the vectors $\vec{Z}^{(1)}$ and $\vec{Z}^{(3)}$. We can also introduce the vectors $\vec{Z}_{**}^{(1)}$ and $\vec{Z}_{**}^{(3)}$ on the basis of integral equations similar to (3.32) for the regions S_2 and S_4 symmetrical relative to the imaginary axis to the regions S_1 and S_3 (these solutions grow exponentially as $z \to \infty$, $\operatorname{Re} s < 0$). Then, it is obvious that

$$\vec{Z}^{(1)} = \alpha_1 \vec{Z}_{**}^{(1)} + \alpha_2 \vec{Z}^{(2)} + \alpha_3 \vec{Z}_{**}^{(3)} + \alpha_4 \vec{Z}^{(4)}. \tag{3.37}$$

If z is kept fixed, then all four terms must be taken into account in formula (3.37) and not only the "growing" terms $\alpha_1 \vec{Z}_{**}^{(1)}$ and $\alpha_3 \vec{Z}_{**}^{(3)}$. In order to determine the coefficients α_j (more accurately, the principal terms of their asymptotic expansions), we can use the conventional methods, in particular, the asymptotic behavior in the neighborhood of the reversal points.

§ 4. Geometry of Rays and Wave Fronts

Beginning with this section, we will consider the Lamb problem for the case of decreasing velocities $v_p(z)$ and $v_s(z)$ requiring that $n_p'(z) > 0$, $n_s'(z) > 0$ for $z \geq 0$ (this case is analogous to that considered in [7]). In this case, regions of geometrical shadow for the longitudinal and transverse waves arise in the half-space $z > 0$.

We will require the equations for the rays and wave fronts of the various waves in the xz plane for $x \geq 0$.

The ray emerging from the origin at an angle $0 \leq \alpha < \pi/2$ to the Oz axis and propagating with a velocity $v_p(z)$ with $z > 0$ will reach a depth z in time

$$\tau_p(\alpha,\ z) = \int_0^z \frac{n_p^2(z')\, dz'}{\sqrt{n_p^2(z') - n_{p0}^2 \sin^2 \alpha}} \qquad (n_{p0} \equiv n_p(0)). \tag{4.1}$$

The ordinary wavefront (OF) of the longitudinal wave (arc P_1P_0 in Fig. 2) corresponds to such rays.

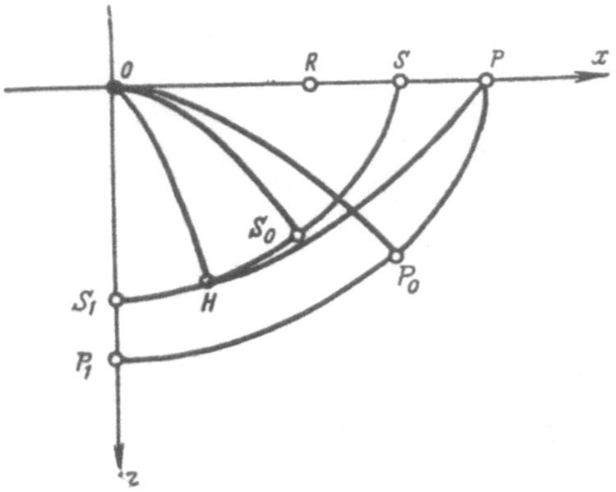

Fig. 2.

For rays that have passed along the boundary $z = 0$ a distance $\delta > 0$ (δ is the slip arc) with a velocity $v_p(0)$ and are propagating with a velocity $v_p(z)$ with $z > 0$, we have

$$\tau_p(\delta, z) = \delta n_{p0} + \int_0^z \frac{n_p^2(z')\,dz'}{\sqrt{n_p^2(z') - n_{p0}^2}}. \tag{4.2}$$

The slip front (SF) of the longitudinal wave (arc P_0P in Fig. 2) corresponds to these rays.

In the case of rays moving in $z \geq 0$ with a velocity $v_s(z)$, the index p in formulas (4.1) and (4.2) should be everywhere replaced by s. The OF and SF of the transverse wave (arcs S_1S_0 and S_0S in Fig. 2) corresponds to these rays.

In the case of rays that have passed along the boundary $z = 0$ a distance $\Delta > 0$ (Δ is the separation arc) with a velocity $v_p(0)$ and are propagating with velocity $v_p(z)$ when $z > 0$, we have

$$\tau_h(\Delta, z) = \Delta n_{p0} + \int_0^z \frac{n_s^2(z')\,dz'}{\sqrt{n_s^2(z') - n_{p0}^2}}. \tag{4.3}$$

The front of the head wave (arc HP in Fig. 2) corresponds to these rays.

For the Rayleigh wave, we have

$$\tau_R(x) = x v_R^{-1}(0), \quad z = 0, \tag{4.4}$$

where $v_R(z) > 0$ satisfies Rayleigh's equation

$$[1 + m_s^2(z, iv_R)]^2 - 4m_s(z, iv_R)\,m_p(z, iv_R) = 0. \tag{4.5}$$

The front of the Rayleigh wave is given by the point R in Fig. 2.

It is convenient to introduce the quantities

$$\gamma_\nu \equiv t - \tau_\nu, \quad \nu = p, s, h, R, \tag{4.6}$$

which have the significance of distances along a ray from the point of observation to the wave front; before the wave front $\gamma < 0$, behind $\gamma > 0$.

§ 5. The Field in the Neighborhood of the Ordinary Fronts of the Longitudinal and Transverse Waves

Here and in the following we give a detailed investigation of the planar Lamb problem with $x \geq 0$;* the corresponding formulas for the problem with axial symmetry are obtained in an analogous manner and are quoted without derivation in § 9.

The ordinary front of the longitudinal waves propagates in the region

*It is obvious that the field in the planar problem is symmetrical with respect to yz.

$$n_{p0}\int_0^z \frac{dz'}{\sqrt{n_p^2(z')-n_{p0}^2}}-x>0. \tag{5.1}$$

The asymptotic behavior of the displacement field in the vicinity of this front can be calculated with the help of the saddle-point method in the same way as was done in [5].

In the present section, it is convenient to extend the cuts in the s plane from $s = \pm iv_p(z)$, $s = \pm iv_s(z)$, and $s = 0$ parallel to the real axis in the direction of $-\infty$. It is obvious that the solutions $\vec{G}^{(p)}$ and $\vec{G}^{(s)}$ in the right half-plane will not be affected; the region S_1 will not change, region S_3 will expand into the left half-plane so that the segment $L = [i\varepsilon,\ iv_i - i\varepsilon]$ will now lie completely in S_3.

The solution (2.1) can be written as

$$\vec{u}(x, z, t)=\vec{u}^{(p)}(x, z, t)+\vec{u}^{(s)}(x, z, t), \tag{5.2}$$

where

$$\vec{u}^{(p)}=\frac{1}{\pi}\int_0^\infty dk\,\frac{1}{2\pi i}\int_M \begin{pmatrix} G_1^{(p)}\sin kx \\ G_2^{(p)}\cos kx\end{pmatrix}\frac{D_s e^{kts}}{\nu_0\Delta s}\,ds, \tag{5.3}$$

$$\vec{u}^{(s)}=-\frac{1}{\pi}\int_0^\infty dk\,\frac{1}{2\pi i}\int_M \begin{pmatrix} G_1^{(s)}\sin kx \\ G_2^{(s)}\cos kx\end{pmatrix}\frac{D_p e^{kts}}{\nu_0\Delta s}\,ds. \tag{5.4}$$

This decomposition has a useful property. Let $\vec{\hat{G}}^{(p)}$ and $\vec{\hat{G}}^{(s)}$ be any two linearly independent solutions of system (2.3) bounded in the right half-plane. Then, in view of the lemma, we have

$$\vec{G}^{(s)}\equiv\alpha\vec{\hat{G}}^{(s)}+\beta\vec{\hat{G}}^{(p)}=\alpha_1\vec{\hat{G}}^{(s)}+\beta_1\vec{G}^{(p)}, \tag{5.5}$$

so that

$$D_s=\alpha_1\hat{D}_s+\beta_1 D_p,\ E_s=\alpha_1\hat{E}_s+\beta_1 E_p \tag{5.6}$$

(the notation is obvious). From this, we have

$$\Delta(k, s)=\alpha_1(k, s)\hat{\Delta}(k, s), \tag{5.7}$$

so that the decomposition (5.2) is equivalent to the decomposition

$$\vec{u}(x, z, t)=\vec{\hat{u}}^{(p)}(x, z, t)+\vec{\hat{u}}^{(s)}(x, z, t), \tag{5.8}$$

which is obtained from (5.2) by the formal substitution of $\vec{G}^{(s)}$ by $\vec{\hat{G}}^{(s)}$. It is obvious that the decomposition (5.2) possesses the same property of invariance with respect to the choice of the solution $\vec{G}^{(p)}$.

Using the classical asymptotic formulas (§ 3) outside the neighborhoods of the reversal points, we obtain for $z \geq 0$

$$\frac{D_s}{\nu_0\Delta}\vec{G}^{(p)}=-\frac{1}{k}\begin{pmatrix}1\\ m_p\end{pmatrix}\sqrt{\frac{\rho_0 m_{p0}}{\rho m_p}}\cdot\frac{1+m_{s0}^2}{\mu_0 R_0}(1+O(k^{-1}))\,e^{-k\int_0^z m_p d\zeta}, \tag{5.9}$$

$$\frac{D_p}{\nu_0\Delta}\vec{G}^{(s)}=-\frac{2}{k}\binom{m_s}{1}\sqrt{\frac{\rho_0 m_{s0}}{\rho m_s}}\cdot\frac{m_{p0}}{\mu_0 R_0}(1+O(k^{-1}))e^{-k\int\limits_0^z m_s d\zeta}, \tag{5.10}$$

where

$$R\equiv(1+m_s^2)^2-4m_s m_p,\quad f_0\equiv f|_{z=0}.$$

This representation is valid in the s plane with the exception of the lines of zeros of Δ which, as will be seen below, are situated in the left half-space and emerge from the points $\pm iv_p(0)$ and $\pm iv_s(0)$.

The saddle points for the integrands of the Mellin integrals in formulas (5.3) and (5.4) are determined by the conditions

$$t=\int\limits_0^z\frac{n_{p,s}^2(z')\,dz'}{\sqrt{n_{p,s}^2(z')+s^{-2}}}. \tag{5.11}$$

For $\gamma_p\approx 0$ in region (5.1), we have from (5.11)

$$s=\pm\frac{iv_{p0}}{\sin\beta},\quad \beta\underset{\gamma_p\to 0}{\to}\alpha. \tag{5.12}$$

The stationary contours have the same shape as those found in [5]. From the asymptotic representations of $\vec{G}^{(p)}$ and $\vec{G}^{(s)}$ it follows that the expression for $\Delta(k, s)$ does not have zeros in the half-plane Re $s > 0$, at least when $k \gg 1$. Deforming the integration contour in the vicinity of the saddle points of (5.12) to make it coincide with the steepest-descent contour, * we can obtain the following asymptotic formula:

$$\vec{I}_p(z,k,t)\equiv\frac{1}{2\pi i}\int\limits_M\frac{D_s}{\nu_0\Delta}\vec{G}^{(p)}\frac{e^{kts}}{s}\,ds=k^{-\frac{3}{2}}\hat{A}_p(\alpha,z)\,\mathrm{Re}\left\{e^{\frac{\pi i}{4}}\binom{-i}{|m_p|}e^{ik\chi(\beta,z)}(1+O(k^{-1}))\right\}, \tag{5.13}$$

where

$$\hat{A}_p(\alpha,z)=\frac{1}{\sqrt{\pi}\,\mu_0}\sqrt{\frac{\rho_0}{\rho}}\left|\frac{m_{p0}}{m_p}\right|^{1/2}\frac{|1+m_{s0}^2|}{R_0}\cdot\frac{\sin\alpha}{v_{p0}}\sqrt{\frac{2}{|\theta''|}},$$

$$|\theta''|=\int\limits_0^z\frac{n_p^2(z')\,dz'}{|m_p(z',s)|^3};\quad \chi(\beta,z)=n_{p0}\sin\beta\int\limits_0^z\left(n_p^2(z')-n_{p0}^2\sin^2\beta\right)^{-\frac{1}{2}}dz',$$

and all quantities in (5.13) have been calculated with $s = iv_{p0}\operatorname{cosec}\alpha$.

We will assume that for $z \geq 0$ the functions $\lambda(z)$, $\mu(z)$, and $\rho(z)$ are analytic.

The function $\frac{1}{\pi}\int\limits_0^{k_0}\binom{I_{p1}\sin kx}{I_{p2}\cos kx}dk$ is regular in the neighborhood of the surface $\gamma_p = 0$; the term $\vec{u}^{(s)} \equiv 0$ for $\gamma_p \approx 0$, $z \geq \varepsilon > 0$.

---

*Using the classical asymptotic method, we can establish that the function $\Delta(k, s)$ has no zeros for $k \gg 1$ in the left half-plane outside the band $|\mathrm{Im}\, s| \leq v_{p0}$.

From formula (5.13) we obtain directly the asymptotic expression for the displacement field in the vicinity of the OF of the longitudinal wave [$\gamma_p \approx 0,\ z \geq \varepsilon > 0,\ 0 < \varepsilon \leq \alpha \leq (\pi/2) - \varepsilon$]:

$$\vec{u}(x, z, t) = -\begin{pmatrix} 1 \\ |m_p| \end{pmatrix} A_p(\alpha, z)\gamma_+^{1/2}(1 + O(\gamma)), \tag{5.14}$$

where

$$\gamma = \gamma_p, \quad |m_p| = [n_p^2(z) v_{p0}^2 \operatorname{cosec}^2 \alpha - 1]^{1/2},$$

$$A_p(\alpha, z) = \frac{\sqrt{2}}{\pi} \cdot \frac{\cot \alpha}{R_p(\alpha)}\left(\frac{1}{p_0 \sin^2 \alpha} - 2\right)\sqrt{\frac{\rho_0}{\rho}} \cdot \frac{1}{\mu_0 \sqrt{|m_p|}} \cdot \frac{1}{\sqrt{\tau'_\alpha}} > 0,$$

$$R_p(\alpha) = \left(\frac{1}{p_0 \sin^2 \alpha} - 2\right)^2 + 4 \cot \alpha \sqrt{\frac{1}{p_0 \sin^2 \alpha} - 1} > 0,$$

$$\tau'_\alpha = n_{p0}^2 \sin \alpha \cos \alpha \int_0^z n_p^2 (n_p^2 - n_{p0}^2 \sin^2 \alpha)^{-\frac{3}{2}} dz', \quad f_0 \equiv f|_{z=0},$$

$$\gamma_+^\lambda = \begin{cases} \gamma^\lambda, & \gamma > 0, \\ 0, & \gamma \leqslant 0. \end{cases}$$

In the neighborhood of the surface $\gamma_s = 0$ the term $\vec{u}^{(p)}$ is regular, but the term $\vec{u}^{(s)}$ has a singularity. Since, when $\sin \alpha \approx v_{s0}/v_{p0}$ the saddle point for $\vec{u}^{(s)}$ falls into the region $F_p(0)$ in which the classical asymptotic theory cannot be applied to $\Delta(k, s)$, we have to distinguish two separate cases: 1) $\sin^2 \alpha < p_0$ and, 2) $\sin^2 \alpha > p_0$ (the condition $\sin^2 \alpha \neq p_0$ removes from consideration the neighborhoods of the points at which the head and transverse waves meet).

A calculation analogous to that carried out above yields the following expressions for the field in the neighborhood of the OF of the transverse wave:

1) for $0 < \varepsilon \leq \sin \alpha \leq \sqrt{p_0} - \varepsilon$,

$$\vec{u}(x, z, t) = \begin{pmatrix} |m_s| \\ -1 \end{pmatrix} A_s(\alpha, z)\gamma_+^{1/2}(1 + O(\gamma)) + \omega(x, z, t), \tag{5.15}$$

where

$$\gamma = \gamma_s, \quad |m_s| = [n_s^2(z) v_{s0}^2 \operatorname{cosec}^2 \alpha - 1]^{1/2},$$

$$A_s(\alpha, z) = \frac{2\sqrt{2}}{\pi} \cdot \frac{\cot \alpha}{\mu_0 R_s} \sqrt{p_0 \operatorname{cosec}^2 \alpha - 1} \sqrt{\frac{\rho_0}{\rho}} \cdot \frac{1}{\sqrt{|m_s|}} \cdot \frac{1}{\sqrt{\tau'_\alpha}} > 0,$$

$$R_s(\alpha) = (1 - \cot^2 \alpha)^2 + 4 \cot \alpha \sqrt{p_0 \operatorname{cosec}^2 \alpha - 1} > 0,$$

$$\tau'_\alpha = n_{s0}^2 \sin \alpha \cos \alpha \int_0^z n_s^2 (n_s^2 - n_{s0}^2 \sin^2 \alpha)^{-\frac{3}{2}} dz';$$

$\omega(x, z, t)$ is a regular function.

2) for $\sqrt{p_0} + \varepsilon \leq \sin \alpha \leq 1 - \varepsilon$,

$$\vec{u}(x, z, t) = \begin{pmatrix} |m_s| \\ -1 \end{pmatrix} \widetilde{A}_s(\alpha, z)\left[\gamma_+^{1/2} + \varkappa\gamma_-^{1/2}\right](1 + O(\gamma)) + \omega(x, z, t), \tag{5.16}$$

where

$$\varkappa = \frac{a}{b};\ a = (1 - \cot^2\alpha)^2 > 0,\ b = 4\cot\alpha\sqrt{1 - p_0\operatorname{cosec}^2\alpha} > 0,$$

$$\widetilde{A}_s(\alpha, z) = \frac{2}{\pi\mu_0}\cdot\frac{b\cot\alpha}{\widetilde{R}_s}\sqrt{1 - p_0\operatorname{cosec}^2\alpha}\sqrt{\frac{\rho_0}{\rho}}\cdot\frac{1}{\sqrt{|m_s|}}\cdot\frac{1}{\sqrt{\tau'_\alpha}} > 0,$$

$$\widetilde{R}_s = a^2 + b^2;$$

τ'_α is the same expression as in (5.15); $\omega(x, z, t)$ is a regular function;

$$\gamma_-^\lambda = \begin{cases} |\lambda|^\lambda, & \gamma < 0, \\ 0, & \gamma > 0. \end{cases}$$

Formulas (5.14)-(5.16) agree with the formulas of the ray method [9].

§ 6. The Field in the Neighborhood of the Rayleigh Wave Front

Formula (3.15) yields the following representation valid for Re s > 0:

$$k^{-2}\Delta(k, s) = \Delta_0(s) + O(k^{-1}), \tag{6.1}$$

where

$$\Delta_0(s) = p_0 R_0(s),$$

$$R_0(s) \equiv \left[1 + m_s^2(0, s)\right]^2 - 4m_s(0, s)\, m_p(0, s).$$

When the method of introducing cuts indicated in § 5 is used, this representation is also valid in the whole neighborhood of the segment L.

In this section we will impose the following conditions on the velocity $v_s(z)$:

$$v_R(0) < v_i \equiv \inf_{z \in [0, \infty)} v_s(z). \tag{6.2}$$

Since the function $R_0(s)$ has a simple zero at the point $iv_R(0)$ and has no other zeros in the upper half-plane, the equation $\Delta(k, s) = 0$ determines (for sufficiently large k) a unique solution $s_R(k)$ in the neighborhood of the segment L and the following asymptotic expression holds:

$$s_R(k) = iv_R(0) + O(k^{-1}). \tag{6.3}$$

The quantity $s_R(k)$ is pure imaginary since the quantities $\Delta(k, i\sigma)$ and $\Delta_0(i\sigma)$ are real for $0 < \varepsilon \le \sigma \le v_i - \varepsilon$, and $v_R(0)$ is real. In the lower half-plane there is a root $\overline{s_R(k)}$.

In order to calculate the first correction term in formula (6.3), we must set m = 2 in the corresponding formulas of § 3 and to the conditions (3.10) and (3.11) add the conditions $\alpha_2 > 1/2$ and

$$\lambda''', \mu''', \rho''' = O\left(z^{-1-\gamma}\right),\ z \to \infty,\ \gamma > 0. \tag{6.4}$$

This calculation has been carried out in [15]; the result is

$$s_R(k) = i\left[v_R(0) + k^{-1}v_1 + O(k^{-2})\right],$$

where v_1 depends on $\lambda_0^{\prime}$, $\mu_0^{\prime}$, $\rho_0^{\prime}$, and may be either positive or negative [15].

Let us calculate the sum of the residues $\vec{R}(k,t)$ of the integrand in formula (2.1) for $z = 0$ corresponding to the roots s_R and $\bar{s}_R$. We find that

$$\vec{R}(k, t) = -\frac{\hat{A}_R}{k}\begin{pmatrix}1-\frac{r_0}{2}\\ -m_{p0}\end{pmatrix}\operatorname{Re}\left\{e^{i(ktv_{R0}+tv_1)}\left(1+O(k^{-1})\right)\right\}, \tag{6.6}$$

where

$$\hat{A}_R = \frac{1}{2\mu_0 g} > 0,\quad r_0 = \frac{v_{R0}^2}{v_{s0}^2} < 1,\quad v_{R0} \equiv v_R(0),$$

$$g = r_0 - 2 + \frac{m_{p0}}{m_{s0}} + p_0\frac{m_{s0}}{m_{p0}} > 0,\quad m_{p0} = \sqrt{1-p_0 r_0},\quad m_{s0} = \sqrt{1-r_0}\,.$$

Let us deform the integration contour in the s plane for $z = 0$ and $\gamma_R \approx 0$ in such a manner that it intersects the points $\pm s_R$. After integration with respect to k, the residue sum $\vec{R}(k,t)$ will give the principal part of the Rayleigh wave field, while the remaining integral will be a regular function in the neighborhood of $\gamma_R = 0$.

Thus, for the Rayleigh wave field we have for $z = 0$

$$\vec{u}(x, 0, t) = A_R \operatorname{Re}\left\{\begin{pmatrix} i\left(1-\frac{r_0}{2}\right)\\ -m_{p0}\end{pmatrix} e^{itv_1}\ln(\gamma + i0)\right\}\left(1 + O(\gamma)\right) + \omega(x, t), \tag{6.7}$$

where

$$\gamma = \gamma_R,\quad A_R = \frac{1}{4\pi\mu_0 g} > 0,$$
$$\ln(\gamma + i0) = \ln|\gamma| + i\pi\varepsilon(-\gamma);$$

$\omega(x,t)$ is a regular (real) function.

It can be seen that in the propagation of the wave along the axis Ox, the displacement vector does not change in amplitude but only rotates in the xz plane with velocity v_1. This result is also in complete agreement with the result obtained by the ray method [10].

The problem of the stationary surface waves on the boundary of a semi-infinite medium inhomogeneous along the z coordinate has been considered by Zavadskii [17, 18]. Zavadskii has obtained (using another notation) the first two terms of the formal solution of a system of equations of the form of (3.1) for the case when there is a reversal point (not necessarily simple). These formulas were applied by him to the planar problem of surface waves in the case of monotonically decreasing velocities $v_p(z)$ and $v_s(z)$; various special cases of the dependence of λ, μ, and ρ on z were also considered. In particular, Zavadskii has shown that if $\lim_{z\to\infty} v_s(z) < v_{R0}$, then the solution $s_R(k)$ of the dispersion equation $\Delta(k, s) = 0$ has a real part different from zero. This real component leads to the attenuation of a stationary Rayleigh wave as it propagates along the surface. We will not consider here the case when the condition $v_i > v_{R0}$ is violated, because the rigorous justification of the formal expansion must be considered separately in this case.

§ 7. The Field in the Neighborhood of the Slip Fronts of the Longitudinal and Transverse Waves

We can find the asymptotic behavior of $\vec{G}^{(p)}$ and $\vec{G}^{(s)}$ and their derivatives with respect to z at z = 0 in the region $F_p(0)$ with the help of formula (3.30) with m = 1. It is not difficult to establish that if in formula (3.30) we set

$$Ai_1(t) = 2k^{1/6}v(t),\quad Ai_2 = k^{1/6}u(t), \tag{7.1}$$

where u and v are Airy functions as defined by Fok [19], and postmultiply the fundamental matrix obtained by the constant diagonal matrix

$$C(k, s) = \left[\sqrt{m_{p0}}\, e^{-k\int_0^{z_p} m_p d\zeta},\ \sqrt{m_{p0}}\, e^{k\int_0^{z_p} m_p d\zeta}, 1, 1\right], \tag{7.2}$$

then the asymptotic behavior of such a fundamental matrix can be described by classical theory in the half-plane Re s > 0. It follows from this that the asymptotic representation of the function

$$\tilde{\Delta}(k, s) \equiv \Delta(k, s)(m_{p0})^{-\frac{1}{2}} \exp\left(k\int_0^{z_p} m_p d\zeta\right)$$

in the region $F_p(0)$ is given by

$$\tilde{\Delta}(k, s) = k^3 p_0 \left[(1 + m_{s0}^2)^2 y_1(0) + 4m_{s0}k^{-1}y_1'(0) + k^{-1}O(|y_1(0)| + k^{-1}|y_1'(0)|)\right]. \tag{7.3}$$

Therefore, in the neighborhood of $s = iv_{p0}$ the function $\Delta(k, s)$ has the simple zeros (for $k \geq k_0 \gg 1$)

$$s_{pl}(k) = iv_{p0} + e^{\frac{7}{6}\pi i}\alpha_p \varkappa_l k^{-\frac{2}{3}} + \beta_p k^{-1} + O\left(k^{-\frac{4}{3}}\right), \tag{7.4}$$

where

$$\alpha_p = \frac{1}{2n_{p0}}(\ln' n_p^2)^{2/3}|_{z=0} > 0,\quad \beta_p = \frac{2\mu_0^2\sqrt{p_0^{-1} - 1}}{\lambda_0^2 n_{p0}}(\ln' n_p^2)|_{z=0} > 0;$$

$\varkappa_l$ are the zeros of the function $v(-\varkappa)$, $\varkappa_l > 0$, $l = 1, 2, \ldots$

Similarly, from an asymptotic formula of type (3.30) in the region $F_s(0)$ we obtain a series of zeros of the function $\Delta(k, s)$

$$s_{sl}(k) = iv_{s0} + e^{\frac{7}{6}\pi i}\alpha_s \varkappa_l k^{-\frac{2}{3}} - i\beta_s k^{-1} + O\left(k^{-\frac{4}{3}}\right), \tag{7.5}$$

where

$$\alpha_s = \frac{1}{2n_{s0}} (\ln' n_s^2)^{2/3} |_{z=0} > 0, \quad \beta_s = \frac{2}{n_{s0}} \sqrt{1-p_0} (\ln' n_s^2) |_{z=0} > 0.$$

A symmetrical pattern is obtained for the lower half-plane.

We will once again assume that $\lambda(z)$, $\mu(z)$, and $\rho(z)$ are analytic for $z \geq 0$. With $\gamma_p = 0$, $z \geq \varepsilon > 0$, $\delta > 0$, the term $\vec{u}^{(p)}(x, z, t)$ has a singularity and $\vec{u}^{(s)}$ is regular. To calculate the field in the neighborhood of the SF of the longitudinal wave ($\gamma_p = 0$, $\delta > 0$), the inner integral in formula (5.3) can be evaluated by residues, the main contribution, as usual, being given by the poles of (7.4) nearest the imaginary axis. The outer integral is calculated by the saddle-point method.

The asymptotic representation of the field in the neighborhood of $\gamma_p = 0$ for $\delta \geq \delta_0 > 0$, $z \geq \varepsilon > 0$ is

$$\vec{u}(x, z, t) = -\begin{pmatrix} 1 \\ |m_p| \end{pmatrix} B_p(\delta, z)\, e^{\beta_p n_{p0} \delta} \gamma_+^{3/2} e^{-\frac{c_p \delta^{3/2}}{\sqrt{\gamma}}} (1 + O(\sqrt{\gamma})), \tag{7.6}$$

where

$$\gamma = \gamma_p, \quad |m_p| = \sqrt{n_p^2(z) v_{p0}^2 - 1},$$

$$B_p(\delta, z) = \frac{1}{8} \sqrt{\frac{3}{\pi}} \sqrt{\frac{\rho_0}{\rho}} \left(\frac{6}{\varkappa_1 \delta}\right)^2 \frac{1}{v'(-\varkappa_1)\lambda_0 \sqrt[4]{n_p^2 - n_{p0}^2}\, \sqrt{(n_p^2)_0'}} > 0,$$

$$c_p = 2\sqrt{n_{p0}} \left(\frac{\varkappa_1}{6}\right)^{3/2} (\ln' n_p^2)|_{z=0} > 0.$$

In the neighborhood of the SF of the transverse wave ($\gamma_s = 0, \delta > 0$, the term $\vec{u}^{(p)}$ is regular and the term $\vec{u}^{(s)}$ has a singularity at $\gamma_s = 0$.

In calculating this singularity [by the method of residues at the points (7.5)], we encounter the following phenomenon. Although the term $\beta_p k^{-1}$ in formula (7.4) is real, the term $-i\beta_s k^{-1}$ in formula (7.5) gives a complex factor $\exp\{-i\beta_s n_{s0}(\gamma v_{s0} + \delta)\}$ in the integrand for the integration with respect to k. We then obtain the following formula for the field in the neighborhood of the SF of the transverse wave (with $\delta \geq \delta_0 > 0$, $z \geq \varepsilon > 0$):

$$\vec{u}(x, z, t) = -\begin{pmatrix} |m_s| \\ -1 \end{pmatrix} B_s(z)\, \mathrm{Re}\left\{e^{-i\beta_s n_{s0}(\gamma v_{s0} + \delta)} F_{-\frac{7}{2}}(\gamma, \delta)(1 + O(\sqrt{\gamma}))\right\} + \omega(x, z, t), \tag{7.7}$$

where

$$\gamma = \gamma_s, \quad |m_s| = \sqrt{n_s^2(z) v_{s0}^2 - 1},$$

$$B_s(z) = \frac{3}{2\pi} \sqrt{\frac{\rho_0}{\rho}} \cdot \frac{\sqrt{1-p_0}}{\mu_0 v'(-\varkappa_1)} \cdot \frac{\sqrt{n_{s0}}}{\sqrt[4]{n_s^2 - n_{s0}^2}} (\ln' n_s^2)^{5/6}|_{z=0} > 0,$$

$$F_\lambda(\gamma, \delta) = \int_{L(\gamma)} \xi^\lambda \exp\{v_{s0}\gamma\xi^3 - (v_{s0}\gamma + \delta)\alpha_s \varkappa_1 n_{s0} \xi\}\, d\xi;$$

$\omega(x, z, t)$ is a regular function; $L(\gamma)$ denotes a contour going from any fixed point $\xi_0 > 0$ to the point $\infty \exp[(\pi i/3)\varepsilon(\gamma)]$.

The expression $\widetilde{F} \equiv \operatorname{Re}\left\{e^{-i\beta_s n_{s0}(\delta+\gamma v_{s0})} F_{-\frac{7}{2}}(\delta, \gamma)\right\}$ represents a nonanalytic function of the real variable γ (with a fixed $\delta > 0$). The asymptotic representation of the function $\widetilde{F}(\gamma, \delta)$ as $\gamma \to 0$ can be found for particular values of the quantity $\widetilde{\delta} \equiv \delta + \gamma v_{s0}$. Thus, if $\beta_s n_{s0}\widetilde{\delta} = (\pi/2) + k\pi$ $(k = 0, \pm 1, \ldots)$, then

$$\widetilde{F}(\gamma, \delta) = \frac{1}{2} \int_{\infty e^{-\frac{\pi i}{3}}}^{\infty e^{\frac{\pi i}{3}}} \xi^{-\frac{7}{4}} e^{\gamma v_{s0}\xi^3 - \alpha_s \varkappa_1 n_{s0} \widetilde{\delta} \xi} \, d\xi \cdot \varepsilon(\gamma). \tag{7.8}$$

This integral can be easily calculated by the saddle-point method and yields a formula analogous to (7.6).

It appears that this type of behavior of the field in the neighborhood of the SF of the transverse wave is associated with the fact that the shadow region for the transverse wave is situated beyond the limiting ray of the head wave, i.e., beyond the singular point of the front.

§ 8. The Field in the Neighborhood of the Front of the Head Wave

The field in the neighborhood of the front of the head wave can be calculated in the same way as was done above. The Mellin integral in expression (5.4) for $\vec{u}^{(s)}$ is found by the method of residues at the poles of (7.4) closest to the imaginary axis and the integration with respect to the parameter k is carried out by the saddle-point method. [The term $\vec{u}^{(p)}$ is regular in the vicinity of $\gamma_h = 0$, $\Delta > 0$.] The following formula for the field in the vicinity of the front $\gamma_h = 0$ of the head wave is obtained ($\Delta \geq \Delta_0 > 0$; $z \geq \varepsilon > 0$):

$$\vec{u}(x, z, t) = -\begin{pmatrix} |m_s| \\ -1 \end{pmatrix} B_h(\Delta, z) e^{\beta_p n_{p0} \Delta} \gamma_+^{7/4} e^{-\frac{c_p \Delta^{3/2}}{\sqrt{\gamma}}} (1 + O(\sqrt{\gamma})) + \omega(x, z, t), \tag{8.1}$$

where

$$\gamma = \gamma_h, \ |m_s| = \sqrt{n_s^2(z) v_{p0}^2 - 1}, \ f_0 \equiv f|_{z=0},$$

$$B_h(\Delta, z) = \frac{1}{2} \sqrt{\frac{3}{\pi}} \left(\frac{6}{\varkappa_1 \Delta}\right)^{9/4} \sqrt{\frac{\rho_0}{\rho}} \cdot \frac{v_{p0}^{5/4} \mu_0}{\lambda_0^2} \, \frac{\sqrt{p_0^{-1} - 1}}{\sqrt[4]{n_s^2 - n_{s0}^2}} \cdot \frac{1}{\sqrt{(\ln' n_p^2)_0}} > 0;$$

$\omega(x, z, t)$ is a regular function; c_p is the same quantity as appears in formula (7.6).

It can be seen that the singularity on the front of the head wave is weaker than that on the SF of the longitudinal wave, which was to be expected.

§ 9. Formulas for the Axially Symmetric Lamb Problem

Let us write down the asymptotic formulas for the field in the case of the axially symmetric Lamb problem in the neighborhood of the following wave fronts.

1) <u>OF of the Longitudinal Wave</u>

$$\vec{u}(r,\ z,\ t) = -\begin{pmatrix} 1 \\ |m_p| \end{pmatrix} \frac{1}{\sqrt{r}} A_p^0(\alpha,\ z)\,\varepsilon(\gamma)\,(1 + O(\gamma)), \tag{9.1}$$

where

$$\gamma = \gamma_p,\quad A_p^0(\alpha,\ z) = \pi \sqrt{\frac{1}{2} n_{p0} \sin\alpha}\ A_p(\alpha,\ z);$$

2) <u>OF of the Transverse Wave</u>

a. for $0 < \varepsilon \leq \sin\alpha \leq \sqrt{p_0} - \varepsilon$,

$$\vec{u}(r,\ z,\ t) = \begin{pmatrix} |m_s| \\ -1 \end{pmatrix} \frac{1}{\sqrt{r}} A_s^0(\alpha,\ z)\,\varepsilon(\gamma)(1 + O(\gamma)) + \omega(r,\ z,\ t), \tag{9.2}$$

where

$$\gamma = \gamma_s,\quad A_s^0(\alpha,\ z) = \pi \sqrt{\frac{1}{2} n_{s0} \sin\alpha}\ A_s(\alpha,\ z),$$

b. for $\sqrt{p_0} + \varepsilon \leq \sin\alpha \leq 1 - \varepsilon$,

$$\vec{u}(r,\ z\ t) = \begin{pmatrix} |m_s| \\ -1 \end{pmatrix} \frac{1}{\sqrt{r}} \tilde{A}_s^0(\alpha,\ z)\,[\varkappa \ln|\gamma| + \pi\varepsilon(\gamma)] + \omega(r,\ z,\ t), \tag{9.3}$$

where

$$\gamma = \gamma_s,\quad \widetilde{A}_s^0(\alpha,\ z) = \sqrt{\frac{1}{2} n_{s0} \sin\alpha}\ \widetilde{A}_s(\alpha,\ z);$$

3) <u>Front of the Rayleigh Wave</u>

$$\vec{u}(r,\ 0,\ t) = -\frac{1}{\sqrt{r}} A_R^0 \operatorname{Re}\left\{\begin{pmatrix} i\left(1 - \frac{r_0}{2}\right) \\ m_{p0} \end{pmatrix} e^{itv_1}\left(\gamma_-^{-\frac{1}{2}} + i\gamma_+^{-\frac{1}{2}}\right)(1 + O(\gamma)\right\} + \omega(r,\ t), \tag{9.4}$$

where

$$\gamma = \gamma_R,\quad A_R^0 = \pi\sqrt{2}\,A_R;$$

4) <u>SF of the Longitudinal Wave</u>

$$\vec{u}(r,\ z,\ t) = \begin{pmatrix} 1 \\ |m_p| \end{pmatrix} \frac{1}{\sqrt{r}} B_p^0(\delta,\ z)\, e^{\beta_p n_{p0}\delta} \gamma_+^{3/4} e^{-\frac{c_p \delta^{3/2}}{\sqrt{\gamma}}} (1 + O(\sqrt{\gamma})), \tag{9.5}$$

where $\gamma = \gamma_p$,

$$B_p^0(\delta,\ z) = \sqrt{2\pi}\left(\frac{\varkappa_1\delta}{6}\right)^{3/4} n_{p0}^{3/4} \sqrt{(\ln' n_p^2)_0}\ B_p(\delta,\ z);$$

5) <u>SF of the Transverse Wave</u>

$$\vec{u}(r,\ z,\ t) = -\begin{pmatrix} |m_s| \\ -1 \end{pmatrix} \frac{1}{\sqrt{r}} B_s^0(z) \operatorname{Re}\left\{e^{-i\beta_s n_{s0}(\delta + \gamma v_{s0})}\ F_{-2}(\gamma,\ \delta)(1 + O(\sqrt{\gamma}))\right\} + \omega(r,\ z,\ t), \tag{9.6}$$

where

$$\gamma = \gamma_s; \quad B_s^0(z) = \sqrt{2\pi}\, B_s(z);$$

6) Front of the Head Wave

$$\vec{u}(r, z, t) = -\begin{pmatrix} |m_s| \\ -1 \end{pmatrix} \frac{1}{\sqrt{r}} B_h^0(\Delta, z) e^{i\beta_p n_p n_{p0} \Delta} \gamma_+ e^{-\frac{c_p \Delta^{3/2}}{\sqrt{\gamma}}} (1 + O(\sqrt{\gamma})) + \omega(r, z, t), \tag{9.7}$$

where

$$B_h^0(\Delta, z) = \sqrt{2\pi} \left(\frac{\varkappa_1 \Delta}{6}\right)^{3/4} n_{p0}^{3/4} \sqrt{(\ln' n_p^2)_0}\, B_h(\Delta, z).$$

In formulas (9.2)-(9.7), the symbol ω denotes a (real) regular function, while the quantities c_p, A_p, A_s, ..., B_h are the same as those defined for the planar problem.

§ 10. Conclusions

The method used in the present paper to solve the Lamb problem allows us, in particular, to estimate the errors of an approximate method developed by Ben-Menahem [13].* The axially symmetric Lamb problem has been considered in [13] for the equations of elasticity with $\lambda = \mu$, $\mu = \mu(z)$, and $\rho = \rho(z)$. The Ben-Menahem method consists in the introduction of the potentials of the longitudinal and transverse waves (as is usually done in the case of a homogeneous semi-infinite medium) and the rejection of terms containing derivatives of λ, μ, and ρ in the system of equations for the potentials. In this approximation the system decomposes into two wave equations for the potentials which can be solved without difficulty by the method of separation of variables.

Unfortunately, this simple approach leads to significant errors, as can be seen on the basis of the method proposed in the present paper. Indeed, the omission of the matrix K(z) in system (3.1) leads to the loss of the factor $\sqrt{\mu(0)/\mu(z)}$ even in the principal term of the asymptotic expansion of the solution of system (3.1) to say nothing of the subsequent approximations. The approximation leads to the loss of such a factor in the first term of the asymptotic solutions of the Lamb problem for other types of waves. In the case of the Rayleigh wave, moreover, the omission of the smaller terms in the equations of elasticity would have led to an incorrect expression for the quantity v_1 equal to the frequency of the polarizational rotation of the displacement vector (the Rayleigh wave is not considered in [13, 20]). On the other hand, using the method proposed in the present paper, we can calculate any term of the asymptotic expansion of the field.

In conclusion, it should be noted that the method for the investigation of the asymptotic properties of the auxiliary system of ordinary differential equations used in the present paper can be also used in the study of the Lamb problem in the case of increasing or nonmonotonic velocities and other problems where systems of "coupled" equations occur (for example, in electromagnetic wave theory).

The author would like to express his sincere gratitude to his scientific supervisor Professor V.M. Babich and to I.A. Molotkov for their help in writing the paper.

---

* The Ben-Menahem method has also been used by Chekin [20] for studying the head wave of the shadow type which arises in the reflection of an elastic spherical wave from the boundary between homogeneous and inhomogeneous semi-infinite media.

Literature Cited

1. H. Lamb, On the propagation of tremors over the surface of an elastic solid. Phil. Trans. Roy. Soc. London, Ser. A, 203 : 1-42 (1904).
2. V.I. Smirnov and S.L. Sobolev, A new method for the solution of the planar problem of elastic vibrations, Tr. Seismol. Inst., No. 20 (1932).
3. V.I. Smirnov and S.L. Sobolev, The application of a new method of studying elastic vibrations, Tr. Seismol. Inst., No. 29 (1933).
4. G.I. Petrashen', The Lamb problem in the case of an elastic semi-infinite medium, Dokl. Akad. Nauk SSSR, 64(5) : 649-652 (1949).
5. G.I. Petrashen', G.I. Marchuk, and K.I. Ogurtsov, The Lamb problem in the case of a semi-infinite medium, Uch. Zap. Leningr. Gos. Univ., 135 : 71-118 (1950).
6. A.S. Alekseev, The Lamb problem for the wave equation in a linearly inhomogeneous semi-infinite medium, Uch. Zap. Leningr. Gos. Univ., 246 : 167-227 (1958).
7. I.A. Molotkov, Nonstationary propagation of waves in an inhomogeneous medium and the formation of a region of geometrical shadow, Dokl. Akad. Nauk SSSR, 140(3) : 557-559 (1961).
8. I.A. Molotkov and I.V. Mukhina, Nonstationary propagation of waves in an inhomogeneous semi-infinite medium in which the propagation velocity has a minimum, in: Abstracts of Reports Presented to Third Symposium on Wave Diffraction, Izd. Nauka (1964), pp. 84-85.
9. V.M. Babich and A.S. Alekseev, The ray method for the calculation of wave front intensities, Izv. Akad. Nauk SSSR, seriya geofiz., 1 : 17-31 (1958).
10. V.M. Babich and N.Ya. Rusakova, The propagation of Rayleigh waves along the surface of an inhomogeneous elastic body of arbitrary shape, Zh. Vychis. Matem i Mat. Fiz., 2(4) : 652-665 (1962).
11. J.D. Tamarkin, Some general problems of the theory of ordinary linear differential equations, Math. Z., 27 : 1-54 (1928).
12. Y. Sibuya, Sur réduction analytique d'un système d'équations differentielles ordinaires lineaires contenant un paramètre. J. Fac. Sci. Univ. Tokyo, 7 : 527-540 (1958).
13. A. Ben-Menahem, Diffraction of elastic waves from a surface in a heterogeneous medium, Bull. Seism. Soc. Am., 50(1) : 15-33 (1960).
14. E.A. Coddington and N. Levinson, The Theory of Ordinary Differential Equations, McGraw-Hill, New York (1955).
15. A.G. Alenitsyn, Rayleigh waves in an inhomogeneous elastic semi-infinite medium, Prikl. Matem. i Mekh., 27(3) : 547-550 (1963).
16. F.W.J. Olver, Uniform asymptotic expansions of solutions of linear second-order differential equations for large values of a parameter. Phil. Trans. Roy. Soc. London, Ser. A, 250 : 387-409 (1958).
17. V.Yu. Zavadskii, Dispersion velocity and attenuation of Rayleigh waves, in: Annotation of Reports Presented to Second Symposium on Wave Diffraction, Izd. Akad. Nauk SSSR (1962), pp. 23-25.
18. V.Yu. Zavadskii, Asymptotic approximations in the dynamics of an elastic inhomogeneous medium, in: Abstracts of Reports Presented to Third Symposium on Wave Diffraction, Izd. Nauka (1964), pp. 72-74.
19. V.A. Fok, Diffraction of Radio Waves Around the Earth's Surface, Izd. Akad. Nauk SSSR, Moscow (1946).
20. B.S. Chekin, The reflection of a spherical elastic wave from an inhomogeneous semi-infinite medium, Izv. Akad. Nauk SSSR, seriya geofiz., 5 : 711-717 (1964).

STIELTJES DOUBLE-INTEGRAL OPERATORS

M. Sh. Birman and M. Z. Solomyak

§ 1. Introduction

In the present paper we investigate operators in separable Hilbert space **H** given by integrals of the type

$$Q = \iint \varphi(\lambda, \mu)\, dF_\mu T dE_\lambda. \tag{1.1}$$

Here, E_λ and E_μ are two orthogonal expansions of unity, T is a bounded operator, and $\varphi(\lambda, \mu)$ is a complex function. The sets E_λ, E_μ and the operator T are not assumed to commute with one another.

Our aim is to establish the properties of the operator Q that depend on the behavior of the function $\varphi(\lambda, \mu)$ and the properties of the operator T. It is clear that we will have to discuss in detail the question of the meaning we can ascribe to integrals of the form (1.1) and to establish the relationships between the various possible definitions of the integral.

Integrals (1.1) were apparently first encountered in the paper by Daletskii and Krein [1], where they were applied to some problems of the analytical theory of perturbations. Repeated integration was assumed. In fact, let the operator $K(\lambda)$ be defined by

$$K(\lambda) = \left(\int \varphi(\lambda, \mu)\, dF_\mu\right) T.$$

Then, by direct definition, the integral is taken to be

$$Q = \int_a^b K(\lambda)\, dE_\lambda, \tag{1.2}$$

where the integration is understood as the limit (in the norm) of the integral sums of Riemann–Stieltjes. The range of integration in (1.2) is assumed to be finite and the function $K(\lambda)$ to be continuously norm-differentiable.* Then, the integral (1.2) exists and

$$\|Q\| \leqslant \max_{a\leqslant\lambda\leqslant b} \|K(\lambda)\| + (b-a) \max_{a\leqslant\lambda\leqslant b} \|K'(\lambda)\|. \tag{1.3}$$

*As shown in [1], this condition can be replaced by the condition that $K(\lambda)$ has an integrable derivative.

The estimate (1.3) immediately leads to the following estimate for the integral (1.1):

$$\|Q\| \leqslant [\sup|\varphi(\lambda, \mu)| + (b-a)\sup|\varphi'_\lambda(\lambda, \mu)|]\,\|T\|. \tag{1.4}$$

Here, the derivative φ'_λ is assumed to be continuous in λ and equicontinuous in μ. If the variables λ and μ can be interchanged, then the integration in the reverse order will lead to the same result.

In a paper by one of the authors [2], integrals of type (1.1) were encountered in connection with perturbations in a continuous spectrum and scattering theory. In [2], the operator T was assumed to be a kernel one. It was found that in this case estimate (1.4) can be replaced by the simpler estimate

$$\|Q\| \leqslant \sup|\varphi(\lambda, \mu)|\,\mathrm{Sp}\,|T|. \tag{1.5}$$

The integral then automatically acquires meaning for a wider class of functions. In subsequent investigations in scattering theory [3] it became necessary to know under what conditions imposed on the function $\varphi(\lambda, \mu)$, the condition that T is a kernel operator implies that the integral operator (1.1) is a kernel operator or is completely continuous. It is comparatively easy to establish this on the basis of some fairly crude assumptions. However, it appeared to be desirable to study this question in greater detail. It is because of this that the authors embarked on the investigation presented in the present paper. Subsequently it was found that there was also common ground with one aspect of the theory of triangle representations of completely continuous operators. This will be discussed in more detail at a later stage.

From the point of view of integration theory it is natural to define the integral in (1.1) as a double integral and not as a repeated one. On the other hand, the attempt to extend the class of integrable functions $\varphi(\lambda, \mu)$ must obviously be based on the definition of the integral in the sense of Lebesgue and not of Riemann. These aims can be fully achieved when T is a Hilbert–Schmidt operator. It is found in this case that for any bounded Borel function the integral Q exists and is also a Hilbert–Schmidt operator (see § 2 for details). However, for the general case it was necessary to restrict integration to the Riemann–Stieltjes definition. One can conjecture that this is to some extent inevitable and corresponds to the fact that the Stieltjes integral

$$\int_a^b f(t)\,dg(t)$$

can exist when each of the functions $f(t), g(t)$ does not have an unbounded variation. It should also be noted that in some problems we have found useful Kolmogorov's approach to the theory of integration. For example, in § 4 we consider integrals of set functions which allows us to obtain the results we require in the simplest possible manner.

Let us now discuss some of the definitions and concepts that we will use systematically in the following.

The ring of all linear bounded operators in **H** will be denoted by **R** and the subspace of all completely continuous operators by $\mathbf{S}_\infty$. It is well known that any completely continuous operator T in Hilbert space **H** can be represented by

$$T = \sum_n s_n(\cdot, \omega_n)\,\vartheta_n, \tag{1.6}$$

where $\{\omega_n'\}$ and $\{\vartheta_n\}$ are two systems orthonormal in **H**, while the numbers $s_n = s_n(T)$ (the eigenvalues of the positive operator $|T|$) tend monotonically to zero. The condition

$$\|T\|_p = \left[\sum_n |s_n(T)|^p\right]^{1/p} < \infty \quad (p \geqslant 1) \tag{1.7}$$

delimits in $\mathbf{S}_\infty$ a class $\mathbf{S}_p$ which is a Banach space corresponding to the norm (1.7). Each of the classes $\mathbf{S}_p$ ($1 \leq p \leq \infty$) is a two-sided ideal in R and for any X, Y in **R** and $T \in \mathbf{S}_p$ we have

$$\|XTY\|_p \leqslant \|X\| \|T\|_p \|Y\|. \tag{1.8}$$

The general form of a linear functional in $\mathbf{S}_p$ ($1 < p < \infty$) is given by the expression

$$f(T) = \operatorname{Sp} F^*T \, (= \operatorname{Sp} TF^*), \tag{1.9}$$

where F is an arbitrary operator in $\mathbf{S}_{p'}$, $p' + p = p'p$. In addition,

$$|\operatorname{Sp} F^*T| \leqslant \|T\|_p \|F\|_{p'} \tag{1.10}$$

and $\|f\| = \|F\|_{p'}$. In other words, $\mathbf{S}_{p'}$ is adjoint to $\mathbf{S}_p$. Moreover, the adjoint space to $\mathbf{S}_\infty$ is $\mathbf{S}_1$, while for $\mathbf{S}_1$ the adjoint is the whole ring **R**. Linear functionals in $\mathbf{S}_\infty$ and $\mathbf{S}_1$ also have the form of (1.9).

Operators of class $\mathbf{S}_1$ are usually called kernel operators, while operators in $\mathbf{S}_2$ are called Hilbert–Schmidt operators. Class $\mathbf{S}_2$ transforms into a Hilbert space when the scalar product is defined by the formula

$$\langle T_1, T_2 \rangle = \operatorname{Sp} T_2^* T_1.$$

It should also be noted that for any two orthonormal bases $\{f_m\}$ and $\{g_n\}$ in **H** and any operator $T \in \mathbf{S}_2$ the following relation holds:

$$\|T\|_2^2 = \sum_{m,\, n} |(Tf_m, g_n)|^2 = \sum_m \|Tf_m\|^2 = \sum_n \|T^*g_n\|^2. \tag{1.11}$$

The theory of the classes $\mathbf{S}_p$ has been developed in detail in monographs [4] and [5].

In papers devoted to the triangular representation of nonself-adjoint operators ([6-10], also see review [11]) the integral operator of the type

$$M = 2i \int_0^1 E_\lambda T dE_\lambda, \tag{1.12}$$

has been studied in detail.* Here, E_λ is some expansion of unity and the operator T is completely continuous. The integral in (1.2) is to be understood as the uniform limit (in the Shatunov sense) of the summation

* The significance of the integral relation (1.12) lies in the fact that with its help, the Volterra operator M is regenerated with respect to its imaginary component $2iT = M^* - M$. It should be recalled that a Volterra operator is an operator of class $\mathbf{S}_\infty$ whose spectrum consists of only one point, $\lambda = 0$.

$$2i\sum_{k=0}^{n-1} E_{\tilde{\lambda}_k} T(E_{\lambda_{k+1}} - E_{\lambda_k}) \quad \left(\lambda_k \leqslant \tilde{\lambda}_k \leqslant \lambda_{k+1}\right), \tag{1.13}$$

the necessary condition for convergence being

$$(E_{\lambda+0} - E_\lambda) T (E_{\lambda+0} - E_\lambda) = 0 \quad (0 \leqslant \lambda \leqslant 1). \tag{1.14}$$

If (1.14) holds, then as was shown by Brodskii [6], $M \in \mathbf{S}_\infty$ if $T \in \mathbf{S}_1$. Gokhberg and Krein [7, 10] and Matsaev [8, 9] have studied the integral (1.12) as a linear operator which transforms any one of the classes $\mathbf{S}_p$ into another such class (Gokhberg and Krein have proposed that operators acting from one linear operator space into another be called transformers). It was shown that with $1 < p < \infty$ the integral (1.2) is a bounded transformer from $\mathbf{S}_p$ into $\mathbf{S}_p$. In the case $p = 1$ one can only assert that M belongs to class $\mathbf{S}_\Omega$, which is characterized by the condition

$$\sup_n \frac{1}{\ln (n+1)} \sum_{r=1}^{n} s_r(M) < \infty . \tag{1.15}$$

Analogous results were also established for some other normed ideals in $\mathbf{R}$.

The connection of these results with the problem of interest to us will become clear if it is noticed that, even if only formally, the transformer (1.12) can be written as

$$M = 2i \int_0^1 \int_0^1 \chi(\lambda - \mu)\, dE_\mu\, T dE_\lambda, \tag{1.16}$$

where $\chi(\tau) = 0$ when $\tau < 0$, and $\chi(\tau) = 1$ when $\tau \geq 0$. The integral (1.12) can therefore be considered as a double-integral operator of a special kind.

Our aim is to study integrals of the general form (1.1) as transformers in spaces $\mathbf{S}_p$. If such a transformer is bounded from $\mathbf{S}_p$ into $\mathbf{S}_q$, we will call it a transformer of class $(\mathbf{S}_p, \mathbf{S}_q)$. Our attention will be primarily devoted to transformers of classes $(\mathbf{S}_2, \mathbf{S}_2)$, $(\mathbf{S}_1, \mathbf{S}_1)$, $(\mathbf{S}_\infty, \mathbf{S}_\infty)$, and $(\mathbf{R}, \mathbf{R})$. Some intermediate cases can then be considered, as is usual, with the help of the interpolation method. As regards the connection of our work with the investigations devoted to the transformer (1.12), (1.16), the following should be noted. In the proofs of the main theorems about this transformer in papers [6-10], not only was use made of the special character of the integrand, but the proof depends on the special meaning of the transformer itself as one which regenerates the Volterra integral with respect to its imaginary component. The methods of the above papers contribute little to the study of the general case. On the other hand, our methods are badly suited to the study of discontinuous functions and contribute little when applied to the integral (1.16). Only the general theorem of § 2 about transformers of class $(\mathbf{S}_2, \mathbf{S}_2)$ allows us in this case to derive (in another manner) the result of Gokhberg and Krein [7].

Let us briefly summarize the contents of this paper.

In § 2, integrals of type (1.1) are considered as transformers of class $(\mathbf{S}_2, \mathbf{S}_2)$; § 3 is mainly devoted to transformers of class $(\mathbf{S}_1, \mathbf{S}_1)$. Some special questions in the theory of integration are discussed in § 4. The associated analysis is required for establishing the meaning of integrals (1.1) when $T \in \mathbf{R}$. Theorems on transformers of class $(\mathbf{S}_p, \mathbf{S}_q)$ with $q \geq p$ are formulated and proved in § 5. Finally, in § 6 transformers of two special types are considered in greater detail. A preliminary report on the major part of the results of the present paper has appeared in [12].

In conclusion, it should be noted that multivariate singular integrals with characteristics depending on the poles* are essentially integrals of form (1.1). Here, T = I, while both the spectral sets F_μ and E_λ depend on the points λ and μ in multidimensional space. The authors intend to discuss this question in detail in another paper.

§ 2. Transformers of Class $(\mathbf{S}_2, \mathbf{S}_2)$

In the present section we give a definition of integral operators of type (1.1) suitable for any $T \in \mathbf{S}_2$ and we also indicate the conditions for the boundedness of such an integral considered as a transformer of class $(\mathbf{S}_2, \mathbf{S}_2)$. We start by presenting a number of preliminary remarks.

Let $\mathcal{F}_\mu$ be a transformer in Hilbert space $\mathbf{S}_2$ corresponding to a premultiplication by the operator F_μ : $\mathcal{F}_\mu T = F_\mu T$, $T \in \mathbf{S}_2$. In view of (1.8), $\mathcal{F}_\mu$ is a bounded transformer. It can be easily seen that the set of transformers $\mathcal{F}_\mu$ represents the expansion of unity in $\mathbf{S}_2$. In analogous manner we introduce the expansions of unity $\mathcal{E}_\lambda$, corresponding to the postmultiplications of operators in $\mathbf{S}_2$ by E_λ: $\mathcal{E}_\lambda T = TE_\lambda$, $T \in \mathbf{S}_2$. By contrast with the original sets F_μ, E_λ the sets of transformers $\mathcal{F}_\mu, \mathcal{E}_\lambda$ commute

$$\mathcal{F}_\mu \mathcal{E}_\lambda T = \mathcal{E}_\lambda \mathcal{F}_\mu T = F_\mu T E_\lambda \quad (T \in \mathbf{S}_2). \tag{2.1}$$

Relation (2.1) allows us to introduce the spectral measure in the plane of the variables λ, μ. Let $\delta = [a, b)$, $\partial = [c, d)$ be two semi-open intervals and $\Delta = \delta \times \partial$. Let us set

$$G(\Delta) = \mathcal{E}(\delta) \mathcal{F}(\partial) = (\mathcal{E}_b - \mathcal{E}_a)(\mathcal{F}_d - \mathcal{F}_c). \tag{2.2}$$

The additive function of the triangles $G(\Delta)$ can be continued to the orthogonal spectral measure $G(e)$ defined over a certain (G measurable) set class including at least all Borel sets on the plane.

Now, let $\varphi(\lambda, \mu)$ be a Borel function essentially bounded with respect to measure $G(e)$,

$$(G)\text{-}\sup |\varphi(\lambda, \mu)| < \infty. \tag{2.3}$$

From the spectral theory of operators (for example, see [14]), it is known that the integral

$$\Phi = \iint \varphi(\lambda, \mu)\, dG(e) \tag{2.4}$$

is a bounded normal operator (transformer) in Hilbert space $\mathbf{S}_2$, and

$$\|\Phi\| = (G)\text{-}\sup |\varphi(\lambda, \mu)|. \tag{2.5}$$

Next, as is usual, for any X, $T \in \mathbf{S}_2$ we have

$$\langle \Phi T, X \rangle = \iint \varphi(\lambda, \mu)\, d\langle G(e) T, X \rangle,$$

or, what is the same,

$$\operatorname{Sp} X^* \Phi T = \iint \varphi(\lambda, \mu)\, d(\operatorname{Sp} X^* G(e) T). \tag{2.6}$$

Moreover, we have

$$\|\Phi T\|_2^2 = \iint |\varphi(\lambda, \mu)|^2 d\langle G(e) T, T \rangle = \iint |\varphi(\lambda, \mu)|^2 d(\operatorname{Sp} T^* G(e) T). \tag{2.7}$$

---

*See [13] for a discussion of multivariate singular integrals.

It should be noted that if condition (2.3) is not satisfied, then the transformer Φ defined by formula (2.4) is unbounded, while its domain coincides with the set of those $T \in \mathbf{S}_2$ for which the integral on the right-hand side of (2.7) is bounded.

Transformers of form (2.4) corresponding to different functions $\varphi(\lambda, \mu)$ commute with one another. Moreover, the following simple assertion follows from the general "correspondence principle" [14].

1) Let Φ_1, Φ_2 be transformers of type (2.4), corresponding to the functions $\varphi_1(\lambda, \mu)$, $\varphi_2(\lambda, \mu)$. Then, we have

$$\Phi_1\Phi_2 = \Phi_2\Phi_1 = \iint \varphi_1(\lambda, \mu)\varphi_2(\lambda, \mu)\, dG(e). \tag{2.8}$$

2) The transformer Φ^*, conjugate to the transformer (2.4), is defined by

$$\Phi^* = \iint \overline{\varphi(\lambda, \mu)}\, dG(e). \tag{2.9}$$

It should also be noted that the ring of transformers of type (2.4) coincides with the ring of functions of two commuting self-conjugate transformers

$$\mathcal{A} = \int \lambda d\mathcal{E}_\lambda, \quad \mathcal{B} = \int \mu d\mathcal{F}_\mu.$$

The latter correspond to postmultiplication and premultiplication by the self-conjugate operators

$$A = \int \lambda dE_\lambda, \quad B = \int \mu dF_\mu.$$

Let us now consider the character of the convergence of the integral (2.4). As is known, the integral sums of the Lebesgue type converge to the transformer Φ at least strongly in $\mathbf{S}_2$. On the other hand, if the function $\varphi(\lambda, \mu)$ is, for example, continuous, while the measure G(e) is the square of Δ, then the Riemann–Stieltjes integral sums also converge: for any $T \in \mathbf{S}_2$

$$\Phi T = \lim \sum_{k,l} \varphi(\lambda_{kl}, \mu_{kl}) G(\Delta_{kl}) T = \lim \sum_{k,l} \varphi(\lambda_{kl}, \mu_{kl}) F(\partial_l) T E(\delta_k), \tag{2.10}$$

where the triangles $\Delta_{kl} = \delta_k \times \partial_l$ form the decomposition of the square Δ, $(\lambda_{kl}, \mu_{kl}) \in \Delta_{kl}$. The right-hand side of (2.10) obviously corresponds to the formally composed integral sums for the integral (1.1). The above gives us grounds to adopt the following definition.

<u>Definition</u>

Let the operator $T \in \mathbf{S}_2$. The double-integral operator (1.1) is the value of Q of the transformer (2.4) for the operator $T: Q = \Phi T$, or, in detail,

$$Q = \iint \varphi(\lambda, \mu)\, dF_\mu T dE_\lambda = \iint \varphi(\lambda, \mu)\, dG(e) T. \tag{2.11}$$

As we have already noted, the integral (2.11) can be understood as the $\mathbf{S}_2$-norm limit of the integral sums of the Lebesgue type. If the function $\varphi(\lambda, \mu)$ is continuous and the carrier of the G(e) measure is compact, then the integral (1.1), (2.11) is the $\mathbf{S}_2$-norm limit of the integral sums of type (2.10). The above directly leads us to the following theorem.

Theorem 1

Let $\varphi(\lambda, \mu)$ be a Borel function satisfying condition (2.3). Then, for any $T \in \mathbf{S}_2$ the integral operator (2.11) converges in $\mathbf{S}_2$ norm and defines a transformer Φ of class $(\mathbf{S}_2, \mathbf{S}_2)$ with norm (2.5).

Let us also give the formula for the bilinear form of the operator Q. For this purpose we assume in (2.6) that $X = (\cdot, f)g$. Then, since $\mathrm{Sp}X^*Q = (Qf, g)$ we have

$$(Qf, g) = \iint \varphi(\lambda, \mu)\, d\,(G(e)\,Tf, g). \tag{2.12}$$

It should be noted in connection with formula (2.12) that the complex measure $[G(e)Tf, g]$ is the continuation of the additive triangle function

$$(G(\Delta)\,Tf, g) = (TE(\delta)f, F(\partial)\,g). \tag{2.13}$$

The latter, obviously, has a bounded variation not exceeding $\|T\|_2 \|X\|_2 = \|T\|_2 \|f\|\ \|g\|$. This estimate for the variation can be easily established directly. Then, independently of the preceding considerations, the integral on the right-hand side of (2.12) acquires meaning and is estimated by the quantity (G) $-\sup|\varphi(\lambda, \mu)|\ \|f\|\ \|g\|$. This makes it possible for us to give a "weak" definition of the integral Q in terms of the restricted bilinear functional (2.12). This definition is obviously equivalent to the one given earlier.

Let us now make several remarks concerning the special integral (1.16). Inasmuch as the integrand $\chi(\lambda - \mu)$ assumes only the values 0 and 1, the corresponding transformer is bounded and, moreover, is a projector in $\mathbf{S}_2$. It is also easy to show that if the integral (1.12) exists as the limit of the sums (1.13), then it coincides with the integral (1.16). However, the integral (1.12) exists only when condition (1.14) holds. The meaning of the latter condition consists in the requirement that the set of points on the discontinuity of the integrand (the diagonal $\lambda = \mu$) has a zero G measure. Requirements of this type are unavoidable when the integral is defined in the Riemann sense. As regards the integral (1.16), it always exists, but its value will change in general if the integrand on the diagonal is changed.

In conclusion, it should be noted that in our constructions we have not used anywhere the condition for the completeness of the spectral sets E_λ, $\mathbf{F}_\mu$, i.e., the condition $E(-\infty, +\infty) = F(-\infty, +\infty) = I$. Moreover, it is clear that we could have assumed that the spectral measures $E(\delta)$ and $F(\partial)$ are given not on a line, but on a circle.

§ 3. Transformers of Class $(\mathbf{S}_1, \mathbf{S}_q)$

As we have seen, the study of the transformers of class $(\mathbf{S}_2, \mathbf{S}_2)$ was reduced to the use of the well-known results in the spectral theory of operators. The situation is altogether different in the case of transformers of other classes. Here we are forced to use one or another special method of analysis. In this section we will use the method of integral equations with the help of which we are able to study transformers of type $(\mathbf{S}_1, \mathbf{S}_q)$. We will, of course, assume that condition (2.3) is satisfied, while the integral (1.1) will be understood in the sense of definition (2.11) of the preceding section.

It is obvious that it is sufficient to study integral (1.1) for one-dimensional operators T. Indeed, if for any one-dimensional operator T of the form

$$T = (\cdot, \omega)\,\vartheta \quad (\|\omega\| = \|\vartheta\| = 1) \tag{3.1}$$

we are able to obtain the estimate

$$\|\Phi T\|_q \leqslant C_q, \tag{3.2}$$

where the constant C_q is independent of ω and ϑ, then for an operator T of the general type (1.6) we can directly find the required estimate from (3.2):

$$\|\Phi T\|_q \leqslant C_q \|T\|_1. \tag{3.3}$$

Regarding the inequalities of type (3.2), for $q < 2$, clearly, they can only be obtained with additional restrictions imposed on the function $\varphi(\lambda, \mu)$.

Thus, let the operator T in (2.11) be of the form of (3.1). For any f, $g \in \mathbf{H}$ [the additive triangle function (2.13) in the case under consideration] can be expressed as

$$(G(\Delta) Tf, g) = (E(\delta) f, \omega)(\vartheta, F(\partial) g).$$

In accordance with this, expression (2.12) for the bilinear form of the operator $Q = \Phi T$ becomes

$$(Qf, g) = \iint \varphi(\lambda, \mu)\, d(E_\lambda f, \omega)\, d(F_\mu \vartheta, g). \tag{3.4}$$

Let us now introduce the subspace $\mathbf{H}(E_\lambda; \omega) \subset \mathbf{H}$, which is a closed linear span of elements of the type $E(\delta)\omega$, as well as the subspace $L_2(\sigma)$ of functions that are square integrable with respect to measure σ generated by a nondecreasing function $\sigma(\lambda) = (E_\lambda \omega, \omega)$. As is known [14, 15], there exists an isometric mapping U of subspace $\mathbf{H}(E_\lambda; \omega)$ into $L_2(\sigma)$ under which, for an arbitrary interval δ, we have

$$(E(\delta) f, \omega) = \int_\delta u_f(\lambda)\, d\sigma(\lambda). \tag{3.5}$$

Here $f \in \mathbf{H}(E_\lambda; \omega)$ and $u_f(\lambda) = Uf \in L_2(\sigma)$. In a similar manner we define a substance $\mathbf{H}(F_\mu; \vartheta) \subset \mathbf{H}$ and a space $L_2(\tau)$ corresponding to a nondecreasing function $\tau(\mu) = (F_\mu \vartheta, \vartheta)$. The corresponding isometric mapping V of the subspace $\mathbf{H}(F_\mu; \vartheta)$ into $L_2(\tau)$ also possesses the property that for an arbitrary interval ∂ we have

$$(F(\partial)\vartheta, g) = \int_\partial \overline{v_g(\mu)}\, d\tau(\mu). \tag{3.6}$$

Here, $g \in \mathbf{H}(F_\mu; \vartheta)$ and $v_g(\mu) = Vg \in L_2(\tau)$.

It follows directly from (3.4) that the operator Q maps the subspace $\mathbf{H}(E_\lambda; \vartheta)$ into $\mathbf{H}(F_\mu; \vartheta)$ and translates the space $\mathbf{H} \ominus \mathbf{H}(E_\lambda; \omega)$ into a null space. On the other hand, it follows from (3.4)-(3.6) that for $f \in \mathbf{H}(E_\lambda; \omega)$, $g \in \mathbf{H}(F_\mu; \vartheta)$

$$(Qf, g) = \iint \varphi(\lambda, \mu)\, u_f(\lambda) \overline{v_g(\mu)}\, d\sigma(\lambda)\, d\tau(\mu). \tag{3.7}$$

Let us extend the definition of the operator U(V) to all $\mathbf{H}$, assuming it to be zero for $\mathbf{H} \ominus \mathbf{H}(E_\lambda; \omega)$ [$\mathbf{H} \ominus \mathbf{H}(F_\mu; \vartheta)$]. Relation (3.7) means that $VQU^* = K$, where K is an integral operator from $L_2(\sigma)$ into $L_2(\tau)$ defined by the formula

$$v(\mu) \equiv Ku = \int \varphi(\lambda, \mu)\, u(\lambda)\, d\sigma(\lambda). \tag{3.8}$$

Therefore the following theorem holds.

Theorem 2

Let T be a one-dimensional operator (3.1). Then, the corresponding integral operator (1.1) is expressed by the formula

$$Q = V^* K U, \tag{3.9}$$

where K is the integral operator (3.8) acting from $L_2(\sigma)$ into $L_2(\tau)$, while U and V are partly isometric operators mapping **H** into $L_2(\sigma)$ and $L_2(\tau)$, respectively.

The unitary equivalence of the operators Q and K expressed through formula (3.9) reduces the problem to the study of the operator K. Indeed, from (3.9) it follows that

$$Q^* Q = U^* K^* K U.$$

Since the numbers $s_n^2(Q)$ coincide with the eigenvalues of the operator Q^*Q which is unitary equivalent to the operator K^*K, it is sufficient to study the eigenvalues of the latter. The operator K^*K operates in $L_2(\sigma)$ and is given by the Hermitian kernel

$$\psi(\lambda, \nu) = \int \varphi(\lambda, \mu)\, \overline{\varphi(\nu, \mu)}\, d\tau(\mu). \tag{3.10}$$

Let us denote by z_n the sequence of eigenvalues of the kernel $\psi(\lambda, \nu)$. As we have just seen,

$$s_n(Q) = z_n^{-\frac{1}{2}} \quad (n = 1, 2, \ldots). \tag{3.11}$$

On the other hand, the numbers z_n are the sequence of zeros of the Fredholm denominator $D_\psi(z)$ of the kernel $\psi(\lambda, \nu)$. In studying the behavior of the eigenvalues z_n, we can therefore make use of the well-known method based on the estimation of the growth of the integral function $D_\psi(z)$. If such an estimate has been obtained, then we only have to make use of the known relations between the order of growth of the integral function and the distribution of its zeros [16]. In turn, the coefficients of the power expansion of $D_\psi(z)$ can be estimated from the smoothness of the kernel. The methods for the solution of this problem have been developed sufficiently well (see [17-19]).* We still cannot directly use the results of the above-mentioned papers. This is due to the fact that the result we need must be expressed in terms of the initial kernel $\varphi(\lambda, \mu)$, and not of the kernel (3.10). Moreover, only integrals with respect to the Lebesgue measure were considered in the above papers. However, both of these difficulties can be easily overcome. Firstly, to obtain an estimate of the function $D_\psi(z)$ we can use, not the usual Fredholm expansion, but the formula

$$D_\psi(z) = 1 + \sum_{n=1}^{\infty} \frac{(-1)^n z^n}{(n!)^2} \int \left| \varphi\begin{pmatrix} \lambda_1 & \ldots & \lambda_n \\ \mu_1 & \ldots & \mu_n \end{pmatrix} \right|^2 d\sigma(\lambda_1) \ldots d\sigma(\lambda_n)\, d\tau(\mu_1) \ldots d\tau(\mu_n), \tag{3.12}$$

where, as is usual,

$$\varphi\begin{pmatrix} \lambda_1 & \ldots & \lambda_n \\ \mu_1 & \ldots & \mu_n \end{pmatrix} = \det\{\varphi(\lambda_k, \mu_l)\}_{k, l=1, \ldots, n}. \tag{3.13}$$

Relation (3.12), being a special case of Carleman's formula [20] for the Fredholm denominator of the convolution of two kernels, allows us subsequently to use the usual methods for estimating the determinants (3.13). Secondly, it should be recalled that a considerable portion of the results of [17-19] is based on point and not integral estimates of the determinants (3.13). Because of this, the question as to the measure used for the integration becomes unimportant. It is only necessary that both measures σ and τ be finite (which is obviously so in our case) and that the carrier of one of them be compact. The latter restriction is important, and we have

*It should be noted that the first result of this type is to be found in one of Fredholm's papers [17].

to take it into account. However, it can be replaced by a suitable condition on the behavior of $\varphi(\lambda, \mu)$ at infinity. We will consider this at a later stage.

With the above remarks taken into account, the methods of papers [17-19] allow us to obtain the following lemma. Its proof can be reconstructed without difficulty by the reader if he follows the text of the above-mentioned papers.

Lemma 1

Let $\sigma(\lambda)$ and $\tau(\mu)$ be two nondecreasing functions whose total variation is unity and $[a, b]$ be a finite interval. Let the Borel function $\varphi(\lambda, \mu)$ be bounded and with respect to the variable λ satisfy in the interval $[a, b]$ the Lipschitz condition with index $\alpha > 0$ and a constant independent of μ. Then the Fredholm denominator $D_{\psi}(z)$ of the integral operator

$$\int_a^b \psi(\lambda, \nu)\, u(\lambda)\, d\sigma(\lambda) \tag{3.14}$$

with kernel (3.10) has a growth order not exceeding the number $(1 + 2\alpha)^{-1}$. The eigenvalues z_n of the operator (3.14) for any $q > 2(1 + 2\alpha)^{-1}$ satisfy the condition

$$\sum_{n=1}^{\infty} \left(\frac{1}{\sqrt{z_n}}\right)^q \leqslant C_q < \infty, \tag{3.15}$$

where the constant C_q depends on $\varphi(\lambda, \mu)$ and the interval $[a, b]$, but does not depend on the functions $\sigma(\lambda)$ and $\tau(\mu)$. In particular, if $\alpha > 1/2$, then

$$\sum_{n=1}^{\infty} \frac{1}{\sqrt{z_n}} \leqslant C_1 < \infty.$$

On the other hand, if $\alpha = 1/2$, then, in addition to (3.15), for any $q > 1$ we can assert the estimate

$$\frac{1}{\sqrt{z_n}} \leqslant C n^{-1} \quad (n = 1, 2, \ldots) \tag{3.16}$$

with constant C independent of $\sigma(\lambda)$ and $\tau(\mu)$.

The estimates (3.15) and (3.16) are also valid if in some finite interval $[c, d]$ the function $\varphi(\lambda, \mu) \in \mathrm{Lip}\,\alpha$ with respect to the variable μ with constant independent of λ. In this case, the integral in (3.10) should be considered to extend in $[c, d]$, but in (3.14) we can set $a = -\infty$, and $b = +\infty$.

Comparing the results of the lemma with Theorem 2, we directly come to estimates of type (3.2). Thus, the following theorem holds.

Theorem 3

If a bounded Borel function $\varphi(\lambda, \mu)$ satisfies the condition $\mathrm{Lip}\,\alpha$, $\alpha > 0$, with respect to the variable λ (μ) with a constant independent of $\mu(\lambda)$, and if the spectral set E_λ (F_μ) is constant outside any infinite interval, then the integral operator (2.11) defines a transformer Φ of class $(\mathbf{S}_1, \mathbf{S}_q)$ for any $q > 2(1 + 2\alpha)^{-1}$. In particular, if $\alpha > 1/2$, then Φ belongs to class $(\mathbf{S}_1, \mathbf{S}_1)$.

We will now make a few remarks concerning the theorem just proved.

Note 1. The conditions of Theorem 3 are in some sense exact. The corresponding example will be given in § 6. At the same time, they are far from being necessary ones, an example of which is the transformer (1.16). We will present other sufficient conditions in § 6 for transformers of two special types.

Note 2. If the condition of Theorem 3 is satisfied for $\alpha = \frac{1}{2}$ and the operator T is finite-dimensional, then in view of (3.16) and (3.11) we have

$$s_n(Q) = O\left(\frac{1}{n}\right). \tag{3.17}$$

In the case of arbitrary $T \in \mathbf{S}_1$ the estimate (3.17) is no longer in general correct. However, we can assert that in this case the operator Q belongs to class $\mathbf{S}_\Omega$ introduced by Gokhberg and Krein [21]. This class is a Banach space with respect to the norm given by the expression [compare with (1.15)]

$$\|Q\|_\Omega = \sup_n \left[\sum_{k=1}^{n} s_k(Q) \Big/ \sum_{k=1}^{n} (2k-1)^{-1} \right].$$

Note 3. The case of an infinite interval of integration can be reduced to a finite one with the help of a change of variable. For example, let $\xi = \xi(\lambda)$ by a monotonic function mapping an axis into the finite segment $[\alpha, \beta]$. The operator (1.1) then transforms into the integral

$$\int_{-\infty}^{+\infty} \int_{\alpha}^{\beta} \tilde{\varphi}(\xi, \mu)\, dF_\mu T d\tilde{E}_\xi .$$

Here, $\tilde{E}_\xi = E_\lambda$, $\tilde{\varphi}(\xi, \mu) = \varphi[\lambda(\xi), \mu]$. It remains for us to require that the function $\tilde{\varphi}(\xi, \mu)$ satisfy the condition of Theorem 3. The requirement of the smoothness of the function $\tilde{\varphi}$ at $\xi = \alpha$ and $\xi = \beta$ will then be transformed into conditions at infinity for the function $\varphi(\lambda, \mu)$.

Note 4. In § 2 it was noted that the transformers Φ of the form of (2.11) belonging to class $(\mathbf{S}_2, \mathbf{S}_2)$ form a commutative ring in which operations on transformers correspond to the usual algebraic operations on the functions $\varphi(\lambda, \mu)$. The subset of transformers of this ring belonging to class $(\mathbf{S}_1, \mathbf{S}_1)$ obviously forms a subring. In particular, formula (2.8) may be found useful for establishing whether a given transformer belongs to class $(\mathbf{S}_1, \mathbf{S}_1)$. The above refers equally to all transformers $\Phi \in (\mathbf{S}_p, \mathbf{S}_p)$ for any $p < 2$.

Note 5. If $\Phi \in (\mathbf{S}_1, \mathbf{S}_1)$, then the transformer $\overline{\Phi}$ defined by the formula

$$\overline{\Phi} = \iint \overline{\varphi(\lambda, \mu)}\, dF_\mu T dE_\lambda \tag{3.18}$$

also belongs to class $(\mathbf{S}_1, \mathbf{S}_1)$. Indeed, let us represent $T \in \mathbf{S}_1$ as a sum of one-dimensional operators T_n, $n = 1, 2, \ldots$ in accordance with expansion (1.6) and let us take $Q = \Phi T$, $Q_n = \Phi T_n$, $\overline{Q} = \overline{\Phi} T$, $\overline{Q}_n = \overline{\Phi} T_n$. The operators Q_n and $\overline{Q}_n$ correspond to the integral operators of type (3.8) with complex conjugate kernels. It follows from this that $s_k(\overline{Q}_n) = s_k(Q_n)$, $k = 1, 2, \ldots$ Further, denoting by $\|\Phi\|_{1,1}$ the norm of the transformer Φ of class $(\mathbf{S}_1, \mathbf{S}_1)$, we find

$$\|\overline{Q}\|_1 \leqslant \sum_n \|\overline{Q}_n\|_1 = \sum_n \|Q_n\|_1 \leqslant \|\Phi\|_{1,1} \sum_n \|T_n\|_1 = \|\Phi\|_{1,1} \|T\|_1 .$$

This means that $\overline{\Phi} \in (\mathbf{S}_1, \mathbf{S}_1)$. At the same time, we have proved that $\|\overline{\Phi}\|_{1,1} = \|\Phi\|_{1,1}$.

Let us note in conclusion that the results about transformers of class $(\mathbf{S}_p, \mathbf{S}_q)$ for $p \le q < 2$ can be obtained from Theorems 1 and 3 with the help of interpolation. The corresponding formulation will be given in § 5. The situation is more complicated when $p > 2$. Here, we already encounter difficulties with the definition of integral (1.1). In connection with this, § 4 will be devoted to some questions in the theory of integration. The results of this section may possibly be of independent interest.

§ 4. Subsidiary Information About Stieltjes Integrals

In order to proceed to the study of the properties of transformers (1.1) of classes in $\mathbf{R}$ and $\mathbf{S}_p$, $p > 2$, we will have to ascribe meaning to some Stieltjes-type integrals of scalar and vector functions. In the cases considered by us, it is a characteristic feature that the integrating set function may not have bounded variations in any subdomain of the main domain. In these cases we are forced to make use of the concept of Riemann–Stieltjes integrals (and not Lebesgue–Stieltjes integrals).

The approach to the theory of integration proposed by Kolmogorov (see [22]) is in some respects relevant to our discussion. In particular, we will have to integrate multivalued set functions. Such a generalization allows us, for example, in some cases to attach a meaning to a repeated Stieltjes integral when the corresponding double integral possibly may not exist. On the other hand, the consideration of multivalued set functions makes our arguments more easily understandable.

Stieltjes integrals for functions with an unbounded local variation have been considered in a paper by Kondurar' [23]. He proved the existence of the Stieltjes integral

$$\int_a^b g(t)\, dh(t)$$

under the condition that $g(t) \in \mathrm{Lip}\,\alpha$, $h(t) \in \mathrm{Lip}\,\beta$, and $\alpha + \beta > 1$. Theorem 4 given below is ultimately a theorem of the same kind, even if it refers to a considerably more general situation. It contains the theorem of Kondurar' as a very special case. In view of this, it is necessary to remark that some of Kondurar's arguments have been used by us in the proof of Theorem 4.

Let us assume that in the finite interval $\Pi = [a, b)$ we have defined a multivalued vector function $x(\Delta)$ of (semi-open) intervals $\Delta \subset \Pi$ with values X in Banach space $\mathbf{X}$.

Let Λ be the subdivision of the interval Π into the segments Δ_k,

$$\Delta_k = [t_{k-1}, t_k), \quad a = t_0 < t_1 < \dots < t_m = b.$$

The number $l = l(\Lambda) = \max_k (t_k - t_{k-1})$ we will call the diameter of the subdivision Λ. Let us also agree to call the continuation Λ' of the subdivision Λ an elementary continuation if into each interval Δ belonging to the subdivision Λ there falls no more than one point of the new division.

Definition

If, as $l(\Lambda) \to 0$, there exists a limit (with respect to the norm in $\mathbf{X}$) of all possible integral sums

$$S(\Lambda) = \sum_{k=1}^{m} x(\Delta_k), \tag{4.1}$$

corresponding to arbitrary subdivisions Λ of interval Π, then this limit is called the integral of the function $x(\Delta)$ over the interval Π:

$$J = \int_{\Pi} x(d\Delta). \tag{4.2}$$

It should be noted that from the integrability of the function $x(\Delta)$ in the strong sense it follows, of course, that the function is integrable in the weak sense. More accurately, if $\Omega(\cdot)$ is a linear functional over **X**, then the scalar multivalued function of the intervals $\Omega[x(\Delta)]$ is integrable and

$$\int_{\Pi} \Omega(x(d\Delta)) = \Omega\left(\int_{\Pi} x(d\Delta)\right). \tag{4.3}$$

We will now impose two conditions on the function $x(\Delta)$ which will guarantee the existence of the integral (4.2).

Condition A. If the interval Δ is subdivided into two subintervals Δ', Δ'', then there exist values $x^*(\Delta)$, $x^*(\Delta')$, $x^*(\Delta'')$ of the function $x(\cdot)$, such that

$$x^*(\Delta) = x^*(\Delta') + x^*(\Delta'')$$

["weak additivity" of the function $x(\cdot)$].

Condition B. There exists a nondecreasing function $\eta(\lambda)$ satisfying the Dini condition

$$\int_{0} \eta(\lambda)\lambda^{-1}d\lambda < \infty, \tag{4.4}$$

such that for any two integral sums $S(\Lambda)$, $\tilde{S}(\Lambda)$ corresponding to one and the same subdivision Λ of the interval Π, the following condition is satisfied:

$$\left\| S(\Lambda) - \tilde{S}(\Lambda) \right\| \leqslant \eta(l) \quad (l = l(\Lambda)). \tag{4.5}$$

It should be noted immediately that from condition (4.4) follows the convergence of the series

$$\sigma_0 = \sum_{r=0}^{\infty} \eta((b-a)2^{-r}) < +\infty. \tag{4.6}$$

Theorem 4

If the vector function $x(\Delta)$ satisfies in the interval Π conditions A and B, then the integral (4.2) exists, and

$$\|J\| \leqslant 2\sigma_0 + \sup\|x(\Pi)\|. \tag{4.7}$$

We will divide the proof into several stages. For convenience we take $\Pi = [0, 1)$. If the interval $\Delta \subset \Pi$ belongs to the subdivision Λ, we will write $\Delta \in \Lambda$.

I) Let Λ' be the elementary continuation of the subdivision Λ of the interval Π. For the arbitrary integral sum $S(\Lambda)$ we can find a sum $S_0(\Lambda')$ such that

$$\| S(\Lambda) - S_0(\Lambda') \| \leqslant \eta(\tilde{l}); \tag{4.8}$$

here $\tilde{l}$ is the longest of the lengths $\Delta \in \Lambda$, $\bar{\Delta} \in \Lambda'$.

Indeed, if $\Delta \in \Lambda$, $\Delta \in \Lambda'$, then we can take the same value of $x(\Delta)$ in $S_0(\Lambda')$ as in $S(\Lambda)$. For the remaining $\Delta \in \Lambda$, $\Delta \bar{\in} \Lambda'$, in view of condition A, we can choose values of the function $x(\cdot)$, such that $S_0(\Lambda')$ coincides with one of the sums $S^*(\Lambda)$. It remains for us to estimate the quantity $S^*(\Lambda) - S(\Lambda)$ on the basis of condition B.

It should be noted also that for any sum $S(\Lambda')$,

$$\| S(\Lambda) - S(\Lambda') \| \leqslant \eta(\tilde{l}) + \| S_0(\Lambda') - S(\Lambda') \| \leqslant \eta(\tilde{l}) + \eta(l') \leqslant 2\eta(l). \tag{4.9}$$

II) Let Λ', Λ'' be two alternating subdivisions, i.e., subdivisions such that each interval from Λ' (or Λ'') is intersected by not more than two intervals from Λ'' (or Λ'). Then, for any sums $S(\Lambda')$, $S(\Lambda'')$, we have

$$\| S(\Lambda') - S(\Lambda'') \| \leqslant 3\eta(l^*) \quad (l^* = \max(l', l'')). \tag{4.10}$$

To prove this, we note that the product $\bar{\Lambda}$ of subdivisions Λ', Λ'' is an elementary continuation of each of them. Now, in accordance with (4.8), with a special choice of sums $S'(\bar{\Lambda})$, $S''(\bar{\Lambda})$, we find that

$$\| S(\Lambda') - S(\Lambda'') \| \leqslant \| S(\Lambda') - S'(\bar{\Lambda}) \| + \| S(\Lambda'') - S''(\bar{\Lambda}) \| +$$

$$+ \| S'(\bar{\Lambda}) - S''(\bar{\Lambda}) \| \leqslant \eta(l') + \eta(l'') + \eta(\bar{l}) \leqslant 3\eta(l^*).$$

III) The sequence of sums

$$S_p = \sum_{k=1}^{2^p} x\left(\left[\frac{k-1}{2^p}, \frac{k}{2^p}\right)\right) \tag{4.11}$$

converges as $p \to \infty$ to the limit J, which does not depend on the particular choice of functions $x(\cdot)$ in each of the intervals. Estimate (4.7) holds for J as well as the estimate

$$\| S_p - J \| \leqslant 2 \sum_{r=p}^{\infty} \eta(2^{-r}) = 2\sigma_p. \tag{4.12}$$

Indeed, in accordance with (4.9), $\|S_{p+1} - S_p\| \leq 2\eta(2^{-p})$. This leads to the convergence of the sequence S_p to a limit J, and also implies that estimate (4.12) holds. Using (4.12) with $p = 0$, and taking into account that $S_0 = x(\Pi)$, we find estimate (4.5).

IV) Now let Λ be an arbitrary subdivision of the interval $[0, 1)$ for which $l < 2^{-p}$. Let us estimate the difference between the arbitrary sum $S(\Lambda)$ and a sum S_p of the form of (4.11). The subdivsion generating the sum S_p will be denoted by Λ_0 and we note that each interval $\Delta \in \Lambda_0$ contains at least one point t_k generating the subdivision Λ.

We begin by constructing an auxiliary sequence of subdivisions $\Lambda_0, \Lambda_1, \ldots, \Lambda_\mu$ possessing the following properties:

1. Each of the subdivisions $\Lambda_{\alpha+1}$ is an elementary continuation of Λ_α.
2. If $\Delta \in \Lambda_\alpha$ and $\Delta \in \Lambda_{\alpha+1}$, then $\Delta \in \Lambda_\beta$ for all $\beta \geq \alpha$.
3. If (in agreement with the notation of § I) l_α is the longest of the lengths $\Delta \in \Lambda_\alpha$, $\Delta \in \Lambda_{\alpha+1}$, then $2\tilde{l}_{\alpha+1} \leq \tilde{l}_\alpha$, $\alpha = 0, 1, \ldots, \mu - 1$.
4. The subdivision Λ_μ alternates with the subdivision Λ.

We will construct the sequence Λ_α in the following manner on the basis of the subdivision Λ_0. Let us assume that the subdivision Λ_α has been constructed and that $[u, v)$ is any of its intervals. If it contains only one point t_k, then $[u, v) \in \Lambda_{\alpha+1}$. In the converse case, we choose a new subdivision point u' in $[u, v)$. Namely, if $2\tilde{u} = u + v$ and each of the intervals $[u, \tilde{u})$, $[\tilde{u}, v)$ contains the points t_k, then we assume that $u' = \tilde{u}$. On the other hand, if any one of these intervals does not contain the points t_k, then we choose $u' \in [u, v)$ such that the larger of the intervals $[u, u')$, $[u', v)$ contains only one point t_k. (It should be noted that this interval is not subject to further subdivision as we proceed from $\Lambda_{\alpha+1}$ to $\Lambda_{\alpha+2}$.) The addition of new subdivision points u' described above realizes the transition from Λ_α to $\Lambda_{\alpha+1}$.

It is clear that the sequence of interval subdivisions obtained in this manner possesses properties 1-3. As the result of a finite number of steps we arrive at a subdivision Λ_μ each of whose intervals contains exactly one point t_k. In this way condition 4 will also be satisfied.

We are now in a position to estimate the difference of $S(\Lambda)$ and $S_p = S(\Lambda_0)$. Indeed, from the results of §I with a special choice of the sums $S(\Lambda_\alpha)$ we have

$$\| S(\Lambda_{\alpha+1}) - S(\Lambda_\alpha) \| \leqslant \eta(\tilde{l}_\alpha) \leqslant \eta(2^{-p-\alpha}), \quad \alpha = 0, \ldots, \mu - 1,$$

and, consequently,

$$\| S(\Lambda_\mu) - S_p \| \leqslant \sigma_p.$$

From this result and (4.10) we find that

$$\| S(\Lambda_\mu) - S(\Lambda) \| \leqslant 3\eta(2^{-p}) \leqslant 3\sigma_p, \quad \| S(\Lambda) - S_p \| \leqslant 4\sigma_p.$$

The last inequality together with (4.12) yields for $l < 2^{-p}$,

$$\| S(\Lambda) - J \| \leqslant 6\sigma_p.$$

In this manner, the theorem has been proved.

We will also require one generalization of Theorem 4 to the multidimensional case.

Let Π be an n-dimensional parallelepiped: $a_i \leq t_i < b_i$, $i = 1, \ldots, n$. Parallelepipeds of this form we will call (n-dimensional) intervals. Let Λ_i be any subdivision of the edges $[a_i, b_i)$, $i = 1, \ldots, n$, and let $\Lambda = \Lambda_1 \times \ldots \times \Lambda_n$ be the subdivision of the interval Π generated in this way. In the following, we will only consider such subdivisions of the main interval Π and we will take $l = l(\Lambda) = \max_i l(\Lambda_i)$. Again, let $x(\Delta)$ be a multivalued function of the intervals $\Delta \subset \Pi$ with values in $\mathbf{X}$. We are interested in the question of the existence of the integral of this function over the interval Π. It should be noted that the definitions used by us ensure that most of the "one-dimensional" formulations remain in force. In particular, we understand the integral in the sense of the definition given above.

However, we do not succeed in proving the analog of Theorem 4 under conditions A and B. Namely, condition A of weak additivity is found to be insufficient in the multidimensional case. In this connection we note that condition A was used in Theorem 4 only for the proof of inequality (4.8). The latter could have been introduced into the conditions for the theorem instead of condition A. We will make this approach in the multidimensional case. The somewhat artificial nature of the new condition is balanced by the fact that it can be easily verified in cases of interest to us and also allows us to retain proof scheme of Theorem 4.

Thus, we subject the function $x(\Delta)$ to condition B and the following condition.

Condition C. Let Λ be an arbitrary subdivision of Π and Λ' be any continuation of Λ obtained by means of an elementary continuation Λ_i', of the subdivision Λ_i of the edge $[a_i, b_i)$. Then, for any integral sum $S(\Lambda)$ we can find an integral sum $S_0(\Lambda')$, such that

$$\| S(\Lambda) - S_0(\Lambda') \| \leqslant \eta(\tilde{l}_i). \tag{4.13}$$

Here l_i is the longest of the one-dimensional intervals $\delta \in \Lambda_i$, $\delta \bar{\in} \Lambda_i'$.

Theorem 5

If the vector function $x(\Delta)$ of the n-dimensional interval Δ satisfies conditions B and C, then the integral (4.2) exists and the following estimate is valid for it:*

$$\| J \| \leqslant (n+1)\sigma_0 + \sup \| x(\Pi) \|. \tag{4.14}$$

For simplicity we will prove this theorem for the unit square with $n = 2$. We will follow the scheme of Theorem 4 highlighting the differences between the present case and the case $n = 1$.

I) Inequalities (4.13) are now replaced by inequality (4.8).

II) We will call the subdivisions Λ', Λ'' alternating if the corresponding one-dimensional subdivisions Λ_1', Λ_1'', Λ_2', Λ_2'' are alternating. Arguing in the same way as in the case of Theorem 4, we find for any sums $S(\Lambda')$, $S(\Lambda'')$ the estimate

$$\| S(\Lambda') - S(\Lambda'') \| \leqslant 5\eta(l^*) \quad (l^* = \max(l', l'')). \tag{4.15}$$

III) Instead of the sum (4.9) we will consider the sums

$$S_{p,q} = \sum_{j=1}^{2^p} \sum_{k=1}^{2^q} x\left(\left[\frac{j-1}{2^p}, \frac{j}{2^p}\right) \times \left[\frac{k-1}{2^q}, \frac{k}{2^q}\right)\right). \tag{4.16}$$

From condition C follows the existence of the special sums $S^*_{p+1,q}$, $S^*_{p,q+1}$, such that

$$\| S_{p,q} - S^*_{p+1,q} \| \leqslant \eta(2^{-p}), \quad \| S_{p,q} - S^*_{p,q+1} \| \leqslant \eta(2^{-q}). \tag{4.17}$$

Let us now establish that for any $r > 0$

$$\| S_{p+r,p+r} - S_{p,p} \| \leqslant 3\sigma_p. \tag{4.18}$$

Indeed, with successive applications of inequality (4.17) and a special choice of the sum $S^*_{p+r,q+r}$, we find that

$$\| S^*_{p+r,p+r} - S_{p,p} \| \leqslant 2\sigma_p.$$

It remains to note that in view of condition B,

$$\| S^*_{p+r,p+r} - S_{p+r,p+r} \| \leqslant \eta(2^{-p-r}) \leqslant \sigma_p.$$

* The quantity σ_0 is given by formula (4.6) as before, provided that we take $b - a$ to be the longest of the edges of the parallelepiped Π.

Inequality (4.18) yields the limit

$$J = \lim_{p \to \infty} S_{p,p},$$

as well as the estimate

$$\|J - S_{p,p}\| \leqslant 3\sigma_p. \tag{4.19}$$

The latter inequality with p = 0 leads to the estimate (4.14) for the limit J.

IV) Now let $\Lambda = \Lambda_1 \times \Lambda_2$ be an arbitrary subdivision Π and $l = l(\Lambda) < 2^{-p}$. Let $\Lambda^{(0)} = \Lambda_1^{(0)} \times \Lambda_2^{(0)}$ denote the subdivision corresponding to the sum $S_{p,p}$ of the form (4.16). Let us now construct two sequences of one-dimensional subdivisions

$$\Lambda_1^{(0)}, \ldots, \Lambda_1^{(\mu)}; \quad \Lambda_2^{(0)}, \ldots, \Lambda_2^{(\nu)},$$

which possess properties 1-4 (see IV of the proof of Theorem 4) with respect to subdivisions Λ_1 and Λ_2, respectively. Let us also consider the sequence of subdivisions of the interval Π defined as follows:

$$\Lambda^{(\alpha)} = \Lambda_1^{(\alpha)} \times \Lambda_2^{(0)} \quad (0 \leqslant \alpha \leqslant \mu), \quad \Lambda^{(\alpha)} = \Lambda_1^{(\mu)} \times \Lambda_2^{(\alpha-\mu)} \quad (\mu < \alpha \leqslant \mu + \nu).$$

It is obvious that $\Lambda^{(\mu+\nu)}$ alternates with the subdivision Λ under consideration. Because of condition C, with a special choice of the sum $S(\Lambda^{(\alpha)})$ we have the following inequalities:

$$\|S(\Lambda^{(\alpha+1)}) - S(\Lambda^{(\alpha)})\| \leqslant \eta(2^{-p-\alpha}) \quad (0 \leqslant \alpha \leqslant \mu - 1),$$

$$\|S(\Lambda^{(\alpha+1)}) - S(\Lambda^{(\alpha)})\| \leqslant \eta(2^{-p-\alpha+\mu}) \quad (\mu \leqslant \alpha \leqslant \mu + \nu - 1).$$

From this, it follows that

$$\|S(\Lambda^{(\mu+\nu)}) - S_{p,p}\| \leqslant 2\sigma_p. \tag{4.20}$$

Next, in accordance with (4.15), for sum $S(\Lambda)$ we have

$$\|S(\Lambda) - S(\Lambda^{(\mu+\nu)})\| \leqslant 5\eta(2^{-p}).$$

Together with (4.17) and (4.18) this gives the estimate

$$\|J - S(\Lambda)\| \leqslant 10\sigma_p.$$

The theorem has thus been proved.

Note. In the multidimensional case we only allowed the subdivision of the main parallelepiped Π by systems of planes that were parallel to the edges. Following the above scheme, we can prove Theorem 5 allowing subdivisions into any parallelepipeds whose edges are parallel to the edges of Π. In doing this, of course, we have to require that conditions B and C hold with respect to a more general system of subdivision.

To conclude the present section, we present one consequence of Theorem 4 which we formulate in the shape of the following theorem.

Theorem 6

Let the operator $K(\mu)$ be a function of the argument $\mu \in [a, b)$ and its values be bounded operators in a Banach space **X**; let $K(\mu)$ satisfy the condition Lip α

$$\|K(\mu_2)-K(\mu_1)\|\leqslant L\,|\mu_2-\mu_1|^{\alpha}\quad(\alpha>0). \tag{4.21}$$

Further, let $f(\Delta)$ be an additive vector function of the intervals $\Delta\subset[a,b)$ with values in $\mathbf{X}$ and let there be constants $M>0$ and $\gamma\in(0,\alpha)$ such that for any subdivision Λ of the interval $[a,b)$ into nonintersecting intervals $\Delta_k=[\mu_{k-1},\mu_k)$, $k=1,\dots,m$ the following inequality holds:

$$\sum_{k=1}^{m}\|f(\Delta_k)\|(\mu_k-\mu_{k-1})^{\gamma}\leqslant M. \tag{4.22}$$

Then, the Stieltjes integral

$$J=\int_a^b K(\mu)\,f(d\mu) \tag{4.23}$$

exists, the integral to be understood as the strong limit as $l(\Lambda)\to 0$ of the integral sums of the form

$$\sum_{k=1}^{m}K(\nu_k)f(\Delta_k),\quad \nu_k\in\Delta_k;$$

and the following inequality holds:

$$\|J\|\leqslant 2LM(b-a)^{\alpha-\gamma}(1-2^{\gamma-\alpha})^{-1}+\|f([a,b))\|\sup_{\mu\in[a,b)}\|K(\mu)\|. \tag{4.24}$$

The proof of the theorem reduces to the investigation of the following multivalued vector function $x(\Delta)$ of the intervals $\Delta=[\mu',\mu'')\subset[a,b)$:

$$x(\Delta)=K(\nu)f(\Delta),\quad \nu\in[\mu',\mu'']. \tag{4.25}$$

The existence of integral (4.2) of it is obviously equivalent to the existence of integral (4.23). It is sufficient therefore to check whether function (4.25) satisfies conditions A and B.

The weak additivity of function $x(\Delta)$ is obvious. Further, in order to check inequality (4.5), we note that

$$\begin{aligned}\|\tilde S(\Lambda)-S(\Lambda)\|&\leqslant\sum_{k=1}^{m}\|[K(\tilde\nu_k)-K(\nu_k)]f(\Delta_k)\|\leqslant\\ &\leqslant L\sum_{k=1}^{m}|\tilde\nu_k-\nu_k|^{\alpha}\|f(\Delta_k)\|\leqslant\\ &\leqslant L\sup_k(\mu_k-\mu_{k-1})^{\alpha-\gamma}\sum_{k=1}^{m}(\mu_k-\mu_{k-1})^{\gamma}\|f(\Delta_k)\|\leqslant LM[l(\Lambda)]^{\alpha-\gamma}.\end{aligned}$$

Condition B is therefore fulfilled with $\eta(l)=LMl^{\alpha-\gamma}$. It remains to note that inequality (4.24) is derived directly from (4.7).

The scalar variant of Theorem 6 contains as a special case the theorem of Kondurar' mentioned above. Indeed, if $h(t)\in\mathrm{Lip}\,\beta$, then the interval function $f([t',t''))=h(t'')-h(t')$ satisfies condition (4.22) with $\gamma=1-\beta$. The Stieltjes integral

$$\int_a^b g(t)\,dh(t)$$

therefore exists if $g(t) \in \mathrm{Lip}\,\alpha$ and $\alpha + \beta > 1$.

§ 5. General Theorems on Transformers

We can now proceed to the study of integral operators (1.1) for an arbitrary bounded operator T. The difficulties here are associated with the fact that even with a "weak" definition of the integral operator in terms of the bilinear expression of type (2.12), the corresponding additive triangle function (2.13) cannot be continued with respect to the (complex) measure with $T \overline{\in} \mathbf{S}_2$. The integrals under consideration can, however, be given a meaning on the basis of the results of the preceding section.

Another possibility for defining the integral operator (1.1) with $T \in \mathbf{R}$ consists in an extended interpretation of formula (2.9). Namely, let it be known that the transformer

$$\overline{\Phi} = \iint \overline{\varphi(\lambda, \mu)}\, dF_\mu T dE_\lambda \tag{5.1}$$

belongs to class $(\mathbf{S}_1, \mathbf{S}_1)$. Since the space $\mathbf{R}$ is conjugate to space $\mathbf{S}_1$, the conjugate transformer $\overline{\Phi}^*$ is obviously a bounded transformer from $\mathbf{R}$ to $\mathbf{R}$ and its norm coincides with a norm of $\overline{\Phi}$ in $(\mathbf{S}_1, \mathbf{S}_1)$. By definition, we now take for any $X \in \mathbf{R}$

$$\Phi X = \iint \varphi(\lambda, \mu)\, dF_\mu X dE_\lambda = \overline{\Phi}^* X. \tag{5.2}$$

In other words, we define the integral operator (5.2) as the value in $X \in \mathbf{R}$ of the bounded transformer $\Phi = \overline{\Phi}^*$.

From relation (2.9) it follows directly that for $X \in \mathbf{S}_2$ this definition of integral (5.2) does not contradict that given in §2. Further, on the basis of (5.2), it is easy to verify that transformers Φ of class $(\mathbf{R}, \mathbf{R})$ form a commutative subring in the commutative ring of transformers Φ of class $(\mathbf{S}_2, \mathbf{S}_2)$. Moreover, according to remark 5 to Theorem 3, conditions $\Phi \in (\mathbf{R}, \mathbf{R})$ and $\Phi \in (\mathbf{S}_1, \mathbf{S}_1)$ are equivalent and the norms of transformer Φ in these classes coincide. Let us also note that the contraction of transformer $\Phi \in (\mathbf{R}, \mathbf{R})$ into $\mathbf{S}_\infty$ is a transformer of class $(\mathbf{S}_\infty, \mathbf{S}_\infty)$. This follows from the fact that for a set of finite-dimensional operators X dense in $\mathbf{S}_\infty$, we have $\Phi X \in \mathbf{S}_2 \subset \mathbf{S}_\infty$, in view of Theorem 1.

Analogous considerations are applicable, obviously, when the function $\varphi(\lambda, \mu)$ generates a transformer $\overline{\Phi}$ of class $(\mathbf{S}_1, \mathbf{S}_q)$, $1 < q < 2$. Formula (5.2) then defines a transformer Φ of class $(\mathbf{S}_p, \mathbf{R})$, $p^{-1} = 1 - q^{-1}$. It is easy to see that in fact this transformer belongs to class $(\mathbf{S}_p, \mathbf{S}_\infty)$.

We will consider definition (5.2) to be the fundamental definition of the integral for $X \in \mathbf{R}$ and $X \in \mathbf{S}_p$, $2 < p < \infty$. Theorem 3 can now be considered as a sufficient condition for the existence of the integral (5.2) for such X, as well as the condition that the transformer

$$\Phi = \iint \varphi(\lambda, \mu)\, dF_\mu(\cdot)\, dE_\lambda \tag{5.3}$$

belongs to classes $(\mathbf{R}, \mathbf{R})$ and $(\mathbf{S}_p, \mathbf{S}_\infty)$, $2 < p \leq \infty$. At the same time, under the conditions of Theorem 3, we can attach to the operator ΦX the meaning of a "real" integral with the help of the results of § 4. This is our aim in the present section.

Thus, let us suppose that the conditions of Theorem 3 are satisfied. For definiteness, we will assume that $\varphi(\lambda, \mu) \in \mathrm{Lip}\,\alpha$ with respect to the variable λ. We will show how it is possible to attach a direct meaning to integral (5.2) for $X \in \mathbf{R}$ with $\alpha > 1/2$ and for $X \in \mathbf{S}_p (p^{-1} > 1/2 - \alpha)$ with $\alpha \leq 1/2$. Let us introduce the operator function

$$K(\lambda) = \int \varphi(\lambda, \mu)\, dF_\mu, \tag{5.4}$$

and let us write integral (5.2) formally as a repeated integral

$$Y = \int_a^b K(\lambda)\, X dE_\lambda. \tag{5.5}$$

We will now establish with the help of Theorem 4 that the integral (5.5) exists as the limit in $\mathbf{R}$ of integral sums of the form

$$\sum_{k=1}^{m} K(\nu_k)\, XE(\delta_k) \quad (\nu_k \in \delta_k). \tag{5.6}$$

In other words, we will prove the integrability in $\mathbf{R}$ of the following multivalued function $x(\delta)$ of the intervals $\delta = [\lambda', \lambda'') \subset [a, b)$:

$$x(\delta) = K(\nu)\, XE(\delta) \quad (\nu \in [\lambda', \lambda'']). \tag{5.7}$$

It is obvious that condition A holds for $x(\delta)$. We will verify that condition B is also satisfied. Let us note, first of all, that

$$\|K(\tilde{\lambda}) - K(\lambda)\| \leqslant \sup_\mu |\varphi(\tilde{\lambda}, \mu) - \varphi(\lambda, \mu)| \leqslant L\,|\tilde{\lambda} - \lambda|^\alpha. \tag{5.8}$$

Let us now estimate the difference between two integral sums $S(\Lambda)$, $\tilde{S}(\Lambda)$ of the type (5.6) corresponding to the same subdivision Λ of the interval $[a, b)$ into subintervals $\delta_k = [\lambda_{k-1}, \lambda_k)$, $k = 1, \ldots, m$. For any element $f \in \mathbf{H}$ we have

$$\begin{aligned} \|[S(\Lambda) - \tilde{S}(\Lambda)]\, f\| &= \left\| \sum_{k=1}^{m} [K(\nu_k) - K(\tilde{\nu}_k)]\, XE(\delta_k) f \right\| \leqslant \\ &\leqslant L \sum_{k=1}^{m} |\lambda_k - \lambda_{k-1}|^\alpha\, \|XE(\delta_k) f\|. \end{aligned} \tag{5.9}$$

From (5.9) with $\alpha > 1/2$, we find that

$$\begin{aligned} \|[S(\Lambda) - \tilde{S}(\Lambda)]\, f\| &\leqslant L\,\|X\|\,[l(\Lambda)]^{\alpha - \frac{1}{2}} \sum_{k=1}^{m} |\lambda_k - \lambda_{k-1}|^{\frac{1}{2}}\, \|E(\delta_k) f\| \leqslant \\ &\leqslant L\,(b-a)^{\frac{1}{2}}\,[l(\Lambda)]^{\alpha - \frac{1}{2}}\,\|X\|\,\|f\| \end{aligned}$$

and, consequently, that

$$\|S(\Lambda) - \tilde{S}(\Lambda)\| \leqslant L\,(b-a)^{\frac{1}{2}}\,[l(\Lambda)]^{\alpha - \frac{1}{2}}\,\|X\|. \tag{5.10}$$

Condition (4.5) is thus satisfied when function $\eta(l)$ is of the form $C\,l^{\alpha-1/2}$.

Let us now take $\alpha \le 1/2$, $X \in \mathbf{S}_p$, and $\varepsilon = \alpha + p^{-1} - 1/2 > 0$. We introduce the orthonormalized elements f_k defined by

$$E(\delta_k)f = \|E(\delta_k)f\| f_k, \quad k = 1, \ldots, m.$$

From inequality (5.9) we find that

$$\left\| \left[S(\Lambda) - \tilde{S}(\Lambda) \right] f \right\| \leqslant L\,[l(\Lambda)]^{\varepsilon} \sum_{k=1}^{m} |\lambda_k - \lambda_{k-1}|^{\frac{1}{2}-\frac{1}{p}} \|Xf_k\| \|E(\delta_k)f\|.$$

Let us now use Hölder's inequality with three indices $[1/2 - (1/p)]^{-1}$, p, $1/2$. We then have

$$\left\| \left[S(\Lambda) - \tilde{S}(\Lambda) \right] f \right\| \leqslant L\,[l(\Lambda)]^{\varepsilon} (b-a)^{\frac{1}{2}-\frac{1}{p}} \|f\| \left\{ \sum_{k=1}^{m} \|Xf_k\|^p \right\}^{\frac{1}{p}} \tag{5.11}$$

To estimate the last factor we use the elementary inequality

$$\|Xf\|^p \leqslant \left\| |X|^{\frac{p}{2}} f \right\|^2 \quad (p \geqslant 2,\ \|f\| = 1).$$

Since

$$\sum_{k=1}^{m} \|Xf_k\|^p \leqslant \sum_{k=1}^{m} \left\| |X|^{\frac{p}{2}} f_k \right\|^2 \leqslant \left\| |X|^{\frac{p}{2}} \right\|_2^2 = \|X\|_p^p,$$

inequality (5.11) directly leads to the estimate

$$\left\| S(\Lambda) - \tilde{S}(\Lambda) \right\| \leqslant L\,(b-a)^{\frac{1}{2}-\frac{1}{p}} [l(\Lambda)]^{\varepsilon} \|X\|_p. \tag{5.12}$$

Condition B is thus satisfied and the existence of integral (5.5) proved.

Let us now assume that the domain of integration in (5.2) is a bounded rectangle $\Pi = [a, b) \times [c, d)$, while the function $\varphi(\lambda, \mu)$ satisfies the condition Lip α with respect to each of its variables,

$$\left| \varphi(\tilde{\lambda}, \tilde{\mu}) - \varphi(\lambda, \mu) \right| \leqslant L \left[|\tilde{\lambda} - \lambda|^{\alpha} + |\tilde{\mu} - \mu|^{\alpha} \right]. \tag{5.13}$$

We will show that with these assumptions and the previous restrictions on the operator X, the operator (5.5) can be represented as a double integral

$$Y = \int_a^b \int_c^d \varphi(\lambda, \mu)\, dF_\mu X dE_\lambda, \tag{5.14}$$

understood as the limit in $\mathbf{R}$ of integral sums of the form

$$\sum_{j,h} \varphi(\lambda_{jh}, \mu_{jh}) F(\partial_j) X E(\delta_h). \tag{5.15}$$

The proof reduces to the verification of the conditions for Theorem 5. Let $x(\Delta)$ be a function of the intervals $\Delta = \delta \times \partial \subset \Pi$ defined by the formula

$$x(\Delta)=\varphi(\lambda, \mu) F(\partial) XE(\delta) \quad (\lambda \in \delta,\ \mu \in \partial), \tag{5.16}$$

and let $\Lambda = \Lambda_1 \times \Lambda_2$ be any subdivision of Π into the intervals $\Delta_{jk} = \delta_k \times \partial_j = [\lambda_{k-1}, \lambda_k) \times [\mu_{j-1}, \mu_j)$, $j = 1, \dots, m$; $k = 1, \dots, n$. Let us verify condition B. Let $S(\Lambda)$, $\tilde{S}(\Lambda)$ be any two integral sums of the type of (5.15) corresponding to the subdivision Λ. For arbitrary $f, g \in \mathbf{H}$, we have

$$\left|\left(\left[S(\Lambda)-\tilde{S}(\Lambda)\right] f, g\right)\right| \leqslant \sum_{j,k} \left|\varphi(\lambda_{jk}, \mu_{jk}) - \varphi\left(\tilde{\lambda}_{jk}, \tilde{\mu}_{jk}\right)\right| \times \left|(F(\partial_j) XE(\delta_k) f, g)\right| \leqslant$$

$$\leqslant L \sum_{j,k} |\lambda_k - \lambda_{k-1}|^{\alpha} \left|(F(\partial_j) XE(\delta_k) f, F(\partial_j) g)\right| + L \sum_{j,k} |\mu_j - \mu_{j-1}|^{\alpha} \left|(E(\delta_k) f, E(\delta_k) X^* F(\partial_j) g)\right| = J_1 + J_2.$$

Let us estimate, for example, the first term

$$J_1 \leqslant L \sum_k |\lambda_k - \lambda_{k-1}|^{\alpha} \sum_j \|F(\partial_j) XE(\delta_k) f\| \|F(\partial_j) g\| \leqslant L \sum_k |\lambda_k - \lambda_{k-1}|^{\alpha} \|XE(\delta_k) f\| \|g\|.$$

It now only remains to apply the estimates obtained above for expression (5.9). As a result of this we find that

$$\left\|S(\Lambda) - \tilde{S}(\Lambda)\right\| \leqslant L\, [l(\Lambda)]^{\alpha - \frac{1}{2}} \|X\| \left[(b-a)^{\frac{1}{2}} + (d-c)^{\frac{1}{2}}\right] \tag{5.17}$$

for $\alpha > 1/2$, $X \in \mathbf{R}$, as well as

$$\left\|S(\Lambda) - \tilde{S}(\Lambda)\right\| \leqslant L\, [l(\Lambda)]^{\varepsilon} \|X\|_p \left[(b-a)^{\frac{1}{2} - \frac{1}{p}} + (d-c)^{\frac{1}{2} - \frac{1}{p}}\right] \tag{5.18}$$

for $\alpha \le 1/2$, $X \in \mathbf{S}_p$, $\varepsilon = \alpha + p^{-1} - 1/2 > 0$.

Condition B has been checked. Condition C is verified in the same way. Let Λ_1' be an elementary continuation of Λ_1 and $\Lambda' = \Lambda_1' \times \Lambda_2$ and let the integral sum (5.15) correspond to the subdivision Λ. The special sum $S_0(\Lambda')$ for the subdivision Λ' we construct as follows. If $\Delta_{jk} \in \Lambda, \Lambda'$, we retain the point (λ_{jk}, μ_{jk}) as before. On the other hand, if $\Delta_{jk} = \Delta'_{jk} \cup \Delta''_{jk}$, then in the intervals Δ'_{jk}, Δ''_{jk} we choose points with the previous values of the second coordinate μ_{jk} for the construction of the sum. The difference $S_0(\Lambda') - S(\Lambda)$ can be estimated in the same way as J_1. Thus, if $\alpha > 1/2$, $X \in \mathbf{R}$, then we have

$$\|S_0(\Lambda') - S(\Lambda)\| \leqslant L\, (b-a)^{\frac{1}{2}}\, \tilde{l}_1^{\,\alpha - \frac{1}{2}} \|X\|,$$

while if $\alpha \le 1/2$, $X \in \mathbf{S}_p$, $p^{-1} > 1/2 - \alpha$,

$$\|S_0(\Lambda') - S(\Lambda)\| \leqslant L\, (b-a)^{\frac{1}{2} - \frac{1}{p}}\, \tilde{l}_1^{\,\varepsilon} \|X\|_p.$$

The conditions of Theorem 5 are therefore satisfied and the existence of integral (5.14) proved. It should also be noted that the coincidence of the operators given by integrals (5.5) and (5.14) can be proved in an elementary manner.

Let us now summarize what has been said above. The following theorem holds.

Theorem 7

Let a bounded Borel function $\varphi(\lambda,\mu)$ satisfy the condition Lip α with respect to variable λ with a constant independent of μ and let the set E_λ be constant outside the finite interval $[a, b)$. Then, for any $X \in \mathbf{R}$ with $\alpha > 1/2$ and for any $X \in \mathbf{S}_p$, $p^{-1} > 1/2 - \alpha$ with $\alpha \leq 1/2$, the repeated integral (5.5) exists as the limit in $\mathbf{R}$ of the integral sums (5.5). This integral coincides with the value in X of transformer (5.3) understood in accordance with definition (5.2)

$$Y = \Phi X. \tag{5.19}$$

The above transformer belongs to classes $(\mathbf{R}, \mathbf{R})$ and $(\mathbf{S}_\infty, \mathbf{S}_\infty)$ for $\alpha > 1/2$ and to class $(\mathbf{S}_p, \mathbf{S}_\infty)$ for $\alpha \leq 1/2$, $p^{-1} > 1/2 - \alpha$. A similar assertion is valid when the roles of the variables λ and μ are interchanged.

If both sets E_λ, F_μ are constant outside the finite intervals and the function $\varphi(\lambda, \mu)$ satisfies condition (5.13), then the operator ΦX also coincides with the double integral (5.14). The latter exists as the limit in $\mathbf{R}$ of the integral sums (5.15).

All of the assertions of this theorem have already been proved with the exception of relation (5.19). The validity of this relation is conveniently established somewhat later after the proof of Theorem 8.

Corollary. Let the operator $T \in \mathbf{S}_1$. Under the conditions of the first part of Theorem 7, the following inequality holds:

$$\operatorname{Sp} T^* Y = \int_a^b \operatorname{Sp} \{T^* K(\lambda) X dE_\lambda\}. \tag{5.20}$$

If, in addition, the conditions of the second part of Theorem 7 are satisfied, then we also have

$$\operatorname{Sp} T^* Y = \int_a^b \int_c^d \varphi(\lambda, \mu) \operatorname{Sp} \{T^* dF_\mu X dE_\lambda\}. \tag{5.21}$$

The existence of integrals (5.20) and (5.21) corresponds to the weak convergence in $\mathbf{R}$ of the integrals (5.5) and (5.14). The relations (5.20) and (5.21) themselves are special cases of relation (4.3). Let us note that the notation used here is convenient, although it is not without defects.

In proving Theorem 7 we have obtained estimates which allow us to strengthen to some extent the results of Theorem 3. Namely, we will show that if the conditions of Theorem 3 hold, then the integral (2.11) with $T \in \mathbf{S}_1$ can be understood as the limit of integral sums in the corresponding $\mathbf{S}_q$ $(q < 2)$ and not only in $\mathbf{S}_2$ as is guaranteed by the results of § 2. At the same time, we will obtain an independent proof of Theorem 3.

Let us suppose, for example, that $\varphi(\lambda,\mu)$ and E_λ satisfy the conditions of the first part of Theorem 7. Let us consider integral sums of the form (5.6) constructed for the operator T and let us prove their $\mathbf{S}_q$-norm convergence $[q > 2(1 + 2\alpha)^{-1}]$. For this purpose, we again make use of Theorem 4 and we apply it to the function $x(\delta)$ of the form (5.7), where T replaces X. Condition A is obviously satisfied. Let $S(\Lambda)$, $\tilde{S}(\Lambda)$ be two sums for the subdivision Λ. Let us estimate their difference in $\mathbf{S}_q$. Let X be an arbitrary operator in $\mathbf{S}_p$ $(p^{-1} + q^{-1} = 1)$. Then, we have

$$\operatorname{Sp}\left\{X^*\left[S(\Lambda)-\tilde{S}(\Lambda)\right]\right\}=\operatorname{Sp}\left\{X^*\sum_{k=1}^{m}\left[K(\nu_k)-K(\tilde{\nu}_k)\right]TE(\delta_k)\right\}=$$
$$=\operatorname{Sp}\left\{T\sum_{k=1}^{m}\left(\left[K^*(\nu_k)-K^*(\tilde{\nu}_k)\right]XE(\delta_k)\right)^*\right\},$$
$$\left|\operatorname{Sp}\left\{X^*\left[S(\Lambda)-\tilde{S}(\Lambda)\right]\right\}\right|\leqslant\|T\|_1\left\|\sum_{k=1}^{m}\left[K^*(\nu_k)-K^*(\tilde{\nu}_k)\right]XE(\delta_k)\right\|.$$

We have already estimated the last expression in the proof of Theorem 7 [see inequality (5.3)]. Using the estimates (5.10) and (5.12), we find that

$$\left|\operatorname{Sp}\left\{X^*\left[S(\Lambda)-\tilde{S}(\Lambda)\right]\right\}\right|\leqslant L(b-a)^{\alpha-\tau}l(\Lambda)^{\tau}\|X\|_p\|T\|_1$$

and, consequently, that

$$\left\|S(\Lambda)-\tilde{S}(\Lambda)\right\|_q\leqslant L(b-a)^{\alpha-\tau}l(\Lambda)^{\tau}\|T\|_1.$$

Here, $\tau = \alpha - \frac{1}{2}$ when $\alpha > \frac{1}{2}$ and $\tau = \alpha + p^{-1} - \frac{1}{2}$ when $\alpha \leq \frac{1}{2}$.

Thus, we see that the integral

$$Q=\int_a^b K(\lambda)\,TdE_\lambda \quad (T\in \mathbf{S}_1) \tag{5.22}$$

exists in the sense of the convergence of the integral sums in $\mathbf{S}_q$. It is not difficult to show that the operator Q defined by formula (5.22) coincides with the operator ΦT defined by formula (2.11). Indeed, it is sufficient to prove this coincidence in the case of bilinear forms of the one-dimensional operator T. But the corresponding measure (2.13) for such T is the product of linear measures and the possibility of the reduction of a double integral to a repeated one follows from the Fubini theorem.

Let us now assume that $\varphi(\lambda,\mu)$, E_λ, F_μ satisfy the conditions of the second part of Theorem 7. The double integral (2.11) can now be understood as the limit of the appropriate Riemann–Stieltjes integral sums in $\mathbf{S}_q$. The verification of the conditions of Theorem 5 also reduces here to the evaluation of estimates already established in the proof of Theorem 7.

The results obtained allow us to formulate the following theorem.

Theorem 8

Let $\varphi(\lambda,\mu)$, E_λ satisfy the conditions of the first part of Theorem 7. Then, for any $T\in\mathbf{S}_1$ the integral (2.11) coincides with the repeated integral (5.22) which is to be understood as the limit in $\mathbf{S}_q$ ($q^{-1} < \frac{1}{2} + \alpha$) of the integral sums

$$\sum_{k=1}^{m} K(\nu_k)\,TE(\delta_k) \quad (\nu_k\in\delta_k).$$

An analogous assertion is valid when the variables λ and μ are interchanged. If $\varphi(\lambda,\mu)$, E_λ, F_μ satisfy the conditions of the second part of Theorem 7, then the double integral sums of the form (5.15) constructed for the operator T converge in $\mathbf{S}_q$ to the integral (2.11).

The following corollary also holds as in the case of Theorem 7.

Corollary. Under the conditions of the first part of Theorem 8, the following relation holds:

$$\operatorname{Sp} X^{*}\Phi T = \int_a^b \operatorname{Sp} \{X^{*}K(\lambda)\, T dE_\lambda\}, \tag{5.23}$$

while under the conditions of the second part of this theorem, we have

$$\operatorname{Sp} X^{*}\Phi T = \int_a^b \int_c^d \varphi(\lambda, \mu) \operatorname{Sp} \{X^{*} dF_\mu T dE_\lambda\}.$$

Here X is any operator in **R** when $\alpha > 1/2$ and in $\mathbf{S}_p$, $p^{-1} > 1/2 - \alpha$, when $\alpha \leq 1/2$.

It should now be noted that the validity of relation (5.19) can be easily established from a comparison of the equalities (5.20) and (5.23).

The results obtained above referred to transformers of classes $(\mathbf{S}_2, \mathbf{S}_2)$, $(\mathbf{S}_1, \mathbf{S}_q)$ $(1 \leq q < 2)$, (S_p, S_∞) $(2 < p \leq \infty)$, and **(R, R)**. As was already noted in § 1, the usual procedures associated with the use of the interpolation theorems allows us to investigate also transformers of intermediate classes. As an example, we quote the following result.

Theorem 9

Let the conditions of Theorem 3 with $\alpha > 1/2$ be satisfied. Then, the function $\varphi(\lambda, \mu)$ defines a transformer (5.3) of class $(\mathbf{S}_p, \mathbf{S}_p)$ for any p, $1 \leq p \leq \infty$.

Indeed, with $p = 1$ and $p = \infty$ the required result is contained in Theorems 3 and 7, so that, according to the interpolation theorem of Riesz, it is valid for all p.

Note. The classical interpolation theorem of Riesz concerning operators in L_p can be extended to transformers in classes $\mathbf{S}_p$ without any change in its formulation. Moreover, the proof of this theorem presented, for example, in [24] remains completely in force for classes $\mathbf{S}_p$. It should be noted that such a variant of Riesz's theorem was used by Gokhberg and Krein [10] in their investigation of the special transformer (1.12). In conclusion, we would like to point out that the more refined interpolation theorem of Martsinkevich (see [24]) can also be extended to transformers in $\mathbf{S}_p$ without any difficulty.

§ 6. Some Remarks About Special Transformers

In the present section we will make several additional observations concerning transformers corresponding to functions of the form

$$\varphi(\lambda, \mu) = \xi(\mu - \lambda),$$

$$\varphi(\lambda, \mu) = \frac{\psi(\lambda) - \psi(\mu)}{\lambda - \mu}. \tag{6.1}$$

We begin with the following elementary assertion of a general character.

Lemma 2

Let the function $\varphi(\lambda, \mu)$ be expanded into the bilinear form

$$\varphi(\lambda, \mu) = \sum_{n=1}^{\infty} p_n(\lambda)\, q_n(\mu),$$

and let

$$C=\sum_{n=1}^{\infty} \sup_{\lambda} |p_n(\lambda)| \sup_{\mu} |q_n(\mu)| < \infty.$$

Then the corresponding transformer (5.3) for any $1 \leq p \leq \infty$ belongs to class $(\mathbf{S}_p, \mathbf{S}_p)$, as well as to class $(\mathbf{R}, \mathbf{R})$.

Indeed, let

$$A=\int \lambda dE_\lambda, \quad B=\int \mu dF_\mu.$$

With $1 \leq p \leq 2$ for any $n = 1, 2, \ldots$ we have

$$Q_n = \Phi_n T = \iint p_n(\lambda)\, q_n(\mu)\, dF_\mu T dE_\lambda = q_n(B)\, T p_n(A),$$

$$\| Q_n \|_p \leqslant \| q_n(B) \| \| T \|_p \| p_n(A) \| \leqslant \sup_{\lambda} | p_n(\lambda) | \sup_{\mu} | q_n(\mu) | \| T \|_p,$$

from which on addition we find that

$$\| \Phi T \|_p \leqslant C \| T \|_p \quad (1 \leqslant p \leqslant 2).$$

For $2 < p \leq \infty$ and in $\mathbf{R}$ the lemma is proved by a transformation to conjugate transformers. The lemma has thus been proved.

Let us now consider the transformer

$$\Xi T = \int_0^{2\pi} \int_0^{2\pi} \xi(\mu - \lambda)\, dF_\mu T dE_\lambda,$$

where $\xi(\tau)$ is a 2π periodic function with an absolutely convergent Fourier series

$$\xi(\tau) = \sum_{-\infty}^{+\infty} c_n e^{in\tau}, \quad \sum_{-\infty}^{+\infty} | c_n | < \infty. \tag{6.2}$$

The assertion of the lemma for the transformer Ξ is obviously satisfied. In particular, in view of the well-known Bernstein theorem (see [24]), condition (6.2) is satisfied when the function $\xi(\tau)$ satisfies the condition $\mathrm{Lip}\,\alpha$ with $\alpha > 1/2$. The same result for transformer Ξ also follows from the general theorem 9 in the last case.

Let us now suppose that $\xi(-\tau) = \overline{\xi(\tau)}$, $E_\lambda = F_\lambda$, $T = (\cdot, \omega)\omega$, and that the expansion E_λ of unity and the element ω are such that $(E_\lambda \omega, \omega) = \lambda$, $\lambda \in [0, 2\pi)$. Then, in accordance with Theorem 2, the operator ΞT is unitary equivalent to the integral operator (with a Hermitian kernel)

$$\int_0^{2\pi} \xi(\mu - \lambda)\, u(\lambda)\, d\lambda, \tag{6.3}$$

operating in $L_2(0, 2\pi)$. It is easy to see that the eigenfunctions of operator (6.3) are the functions $u_n(\lambda) = e^{in\lambda}$, while its eigenvalues are the numbers $2\pi c_n$, where c_n are the coefficients of the expansion (6.2). Bernstein has constructed an example of a function $\xi(\tau) \in \mathrm{Lip}\,1/2$ such that its Fourier series does not converge absolutely,

$$\sum_{-\infty}^{+\infty} |c_n| = \infty .$$

Since for $Q = \Xi T$ we now have $s_n(Q) = 2\pi |c_n|$, this example shows that with $\alpha = 1/2$, even for the one-dimensional operator T it is possible that $\Xi T \overline{\in} \mathbf{S}_1$. It follows from this that the conditions of Theorem 3 cannot be significantly improved. Moreover, other examples of the same type (see [24]) allow us to assert that in Theorem 3 we cannot set $q = 2(1 + 2\alpha)^{-1}$ for any $\alpha > 0$. At the same time, we see that Bernstein's theorem and other analogous results about Fourier series are special cases of general theorems about the behavior of the eigenvalues of integral operators.

Let us now consider transformers Ψ of the form

$$\Psi T = \int_a^b \int_a^b \frac{\psi(\lambda) - \psi(\mu)}{\lambda - \mu} dF_\mu T dE_\lambda. \tag{6.4}$$

From Theorem 9 it follows that $\Psi \in (\mathbf{S}_p, \mathbf{S}_p)$ for any p, provided that $\psi'(\lambda) \in \mathrm{Lip}\,\alpha$ $(\alpha > 1/2)$. With the help of Lemma 2 we now obtain a somewhat better result.

Theorem 10

Let the function $\psi'(\lambda)$ satisfy the condition $\mathrm{Lip}\alpha$ for any $\alpha > 0$. Then, transformer (6.4) belongs to classes $(\mathbf{S}_p, \mathbf{S}_p)$, $1 \le p \le \infty$, and $(\mathbf{R}, \mathbf{R})$.

Proof. Without restricting the generality of the case, we can consider the function $\psi(\lambda)$ to be periodic in $[a, b]$. Indeed, this can be attained by continuing $\psi(\lambda)$ over a wider interval without changing its class. The integral (6.4) will not be changed by this procedure, provided that F_μ, E_λ outside $[a, b]$ are taken to be constant. For convenience of notation we take $a = 0$, $b = 2\pi$.

Let us expand the function $\psi(\lambda)$ into a Fourier series

$$\psi(\lambda) = \sum_{n=-\infty}^{+\infty} c_n e^{in\lambda},$$

and take

$$\psi_+(\lambda) = \sum_{n=1}^{\infty} c_n e^{in\lambda}, \quad \psi_-(\lambda) = \sum_{n=1}^{\infty} c_{-n} e^{-in\lambda}.$$

According to Privalov's theorem (for example, see [24]), $\psi'_\pm(\lambda) \in \mathrm{Lip}\,\alpha$. Next, for function (6.1), we obviously have

$$\varphi(\lambda, \mu) = \frac{e^{i\lambda} - e^{i\mu}}{\lambda - \mu} \sum_{n=1}^{\infty} c_n \frac{e^{in\lambda} - e^{in\mu}}{e^{i\lambda} - e^{i\mu}} + \frac{e^{-i\lambda} - e^{-i\mu}}{\lambda - \mu} \sum_{n=1}^{\infty} c_{-n} \frac{e^{-in\lambda} - e^{-in\mu}}{e^{-i\lambda} - e^{-i\mu}} . \tag{6.5}$$

Using the addition and multiplication rules for transformers, we can reduce the investigation to the study of transformers with the kernels

$$\varphi_\pm(\lambda, \mu) = \sum_{n=1}^{\infty} c_{\pm n} \frac{e^{\pm in\lambda} - e^{\pm in\mu}}{e^{\pm i\lambda} - e^{\pm i\mu}} .$$

Indeed, the factors in both sums in (6.5) are analytic functions and, therefore, the transformers corresponding to them belong to the required classes in view of Theorem 9.

For example, let us transform the expression for $\varphi_+(\lambda, \mu)$. We have

$$\varphi_+(\lambda, \mu) = \sum_{n=1}^{\infty} c_n \frac{e^{in\lambda} - e^{in\mu}}{e^{i\lambda} - e^{i\mu}} = \sum_{n=1}^{\infty} c_n \sum_{k=1}^{n} e^{i(k-1)\mu} e^{i(n-k)\lambda} =$$

$$= \sum_{k=1}^{\infty} \sum_{n=k}^{\infty} c_n e^{i(n-k)\lambda} e^{i(k-1)\mu},$$

from which we find that

$$\varphi_+(\lambda, \mu) = \sum_{k=1}^{\infty} \left[\psi_+(\lambda) - \sum_{n=1}^{k-1} c_n e^{in\lambda} \right] e^{-ik\lambda} e^{i(k-1)\mu}. \tag{6.6}$$

In view of the well-known results (see [24]) on the order of approximation of functions by partial sums of their Fourier series, the condition $\psi_+'(\lambda) \in \mathrm{Lip}\,\alpha$ leads to the estimate

$$\sup_{\lambda} \left| \psi_+(\lambda) - \sum_{n=1}^{k-1} c_n e^{in\lambda} \right| = O\left(\frac{\ln k}{k^{1+\alpha}}\right).$$

From this, it follows directly that the expansion (6.6) of the function $\varphi_+(\lambda, \mu)$ satisfies the conditions of Lemma 2. The theorem has been proved.

Let us give one other result close to Theorem 10.

Theorem 11

Let U and V be two unitary operators in **H**, and $\zeta(t)$ be a function on the circle $|t| = 1$ whose derivative $\zeta'(t)$ satisfies the condition $\mathrm{Lip}\,\alpha$ $(\alpha > 0)$. If $T = V - U \in \mathbf{S}_p$ $(1 \le p \le \infty)$, then

$$\zeta(V) - \zeta(U) \in \mathbf{S}_p.$$

Proof. Let E_λ, F_μ be the spectral sets corresponding to the operators U and V, respectively,

$$U = \int_0^{2\pi} e^{i\lambda} dE_\lambda, \quad V = \int_0^{2\pi} e^{i\mu} dF_\mu.$$

Let us introduce the transformer Z,

$$ZT = \int_0^{2\pi} \int_0^{2\pi} \frac{\zeta(e^{i\mu}) - \zeta(e^{i\lambda})}{e^{i\mu} - e^{i\lambda}} dF_\mu T dE_\lambda. \tag{6.7}$$

In view of Theorem 10, it is obviously sufficient to prove that

$$\zeta(V) - \zeta(U) = ZT.$$

Let us denote for brevity the integrand in (6.7) by $\zeta(\lambda, \mu)$. The transformer Z belongs, in particular, to the class $(\mathbf{R}, \mathbf{R})$ and, therefore, $ZT = ZV - ZU$. We will interpret, for example, the operator ZV as the result of the successive application of two transformers Z and V to the unit operator I. Using the rules for operations on transformers, we find that

$$ZT = \int_0^{2\pi}\int_0^{2\pi} \zeta(\lambda, \mu)\, e^{i\mu} dF_\mu dE_\lambda - \int_0^{2\pi}\int_0^{2\pi} \zeta(\lambda, \mu)\, e^{i\lambda} dF_\mu dE_\lambda =$$
$$= \int_0^{2\pi}\int_0^{2\pi} [\zeta(e^{i\mu}) - \zeta(e^{i\lambda})]\, dF_\mu dE_\lambda = \zeta(V) - \zeta(U).$$

The theorem has thus been proved.

An analogous result holds for a function of self-adjoint operators. It should be noted, in conclusion, that the results of Theorem 11 with $p = 1$ are relevant to abstract scattering theory [2, 3] and the theory of the spectral shift function [25, 26].

Literature Cited

1. Yu. L. Daletskii and S. G. Krein, The integration and differentiation of functions of Hermitian operators and applications to perturbation theory, Transactions of the Seminar on Functional Analysis, Voronezh, Vol. 1 (1956), pp. 81-105.
2. M. Sh. Birman, On the existence of wave operators, Izv. Akad. Nauk SSSR, seriya matem., 27(4): 883-906 (1963).
3. M. Sh. Birman, Local criterion of the existence of wave operators, Dokl. Akad. Nauk SSSR, 159(3): 485-488 (1964).
4. R. Schatten, A Theory of Cross Spaces. Princeton, New Jersey (1950).
5. I. Ts. Gokhberg and M. G. Krein, Introduction to the Theory of Linear Nonself-Adjoint Operators. Izd. Nauka (1966).
6. M. S. Brodskii, The triangle representations of completely continuous operators with one spectral point, Usp. Matem. Nauk, Vol. 16, 1(97): 135-141 (1961).
7. I. Ts. Gokhberg and M. G. Krein, Completely continuous operators with a spectrum concentrated at zero, Dokl. Akad. Nauk SSSR, 128(2): 227-230 (1959).
8. V. I. Matsaev, On one class of completely continuous operators, Dokl. Akad. Nauk SSSR, 139(3): 548-551 (1961).
9. V. I. Matsaev, Volterra operators obtained by the perturbation of self-adjoint operators, Dokl. Akad. Nauk SSSR, 139(4): 810-813 (1961).
10. I. Ts. Gokhberg and M. G. Krein, Volterra operators with an imaginary component, Dokl. Akad. Nauk SSSR, 139(4): 779-782 (1961).
11. M. S. Brodskii, I. Ts. Gokhberg, M. G. Krein, and V. I. Matsaev, Some new results in the theory of nonself-adjoint operators, Transactions of the Fourth All-Union Math. Conference, Vol. 2 (1962), pp. 261-271.
12. M. Sh. Birman and M. Z. Solomyak, Stieltjes double-integral operators, Dokl. Akad. Nauk SSSR, 165(6) (1965).
13. S. G. Mikhlin, Multidimensional Singular Integrals and Integral Equations, Fizmatgiz (1962).
14. A. I. Plesner and V. A. Rokhlin, Spectral theory of linear operators, Part II, Usp. Matem. Nauk, Vol. 1, 1(11): 71-191 (1946).
15. N. I. Akhiezer and I. M. Glazman, The Theory of Linear Operators in Hilbert Space, Gostekhizdat (1950).
16. B. Ya. Levin, The Distribution of the Roots of Integral Functions, Gostekhizdat (1956).
17. I. Fredholm, Sur une class d'equations fonctionelles, Acta Math., 27: 365-390 (1903).
18. E. Hille and J. D. Tamarkin, On the characteristic values of linear integral equations, Acta Math., 57: 1-76 (1931).
19. A. O. Gel'fond, On the growth of eigenvalues of homogeneous integral equations, appendix to: W. V. Lovitt, Linear Integral Equations [Russian translation], GITTL, Moscow (1957).

20. T. Carleman, Zur Theorie der linearen Integralgleichungen, Math. Zs. 9(3-4) : 196-217 (1921).
21. I. Ts. Gokhberg and M. G. Krein, On the theory of the triangle representation of nonself-adjoint operators, Dokl. Akad. Nauk SSSR, 137(5) : 1034-1037 (1961).
22. V. I. Smirnov, A Course in Higher Mathematics, Vol. 5, Gostekhizdat (1947).
23. V. T. Kondurar', Sur l'integrale de Stieltjes, Matem. sb., 2(44) : 361-366 (1937).
24. A. Sigmund, Trigonometric Series, Vols. I and II. [Russian translation]. Izd. Mir (1965).
25. M. G. Krein, The trace formula in perturbation theory, Matem. sb., Vol. 33, 75(3) : 597-626 (1953).
26. M. G. Krein, Perturbation determinants and the trace formula for unitary and self-adjoint operators, Dokl. Akad. Nauk SSSR, 144(2) : 268-271 (1962).

THE INVERSE PROBLEM IN THE THEORY OF SEISMIC WAVE PROPAGATION

A. S. Blagoveshchenskii

Introduction

The following two results have been obtained in the present paper.

1) A new method has been developed for the solution of the inverse problem for a string. The inverse problem for the string has been solved by a number of authors [1-3]. These results are appreciably stronger than those obtained in our paper. Therefore, the first part of the present paper is of purely methodological interest. In all of the preceding papers the inverse problem for the string was solved with the help of spectral methods, while in our paper we have used purely local considerations.

2) The problem of the calculation of elastic constants for a laminarly inhomogeneous semi-infinite medium from given initial and "redundant" boundary conditions has been considered. The formulation of the problem and some results are to be found in [4]. In our paper, the problem is reduced to the successive solution of three inverse problems for a string with the help of a simple device. Two of these are identical with the problem considered in the first part of our paper and can be solved independently of one another. The third has to be solved separately. Its formulation depends on the solutions of the first two problems.

As a result, we were able to prove the uniqueness of the solution of the inverse seismic problem at finite depth with some additional restrictions imposed on $\sigma \equiv \mu/\lambda$ (σ is obtained as the result of the solution of the first two problems).

§ 1. The Inverse Problem for the String Equation

Let the function u(x,t) with x > 0 satisfy an equation of the form

$$u_{tt} = C^2_{(x)} u_{xx}. \tag{1}$$

The initial and boundary conditions on the function u are given as

$$u|_{t<0} \equiv 0, \tag{2}$$

$$u_x|_{x=0} = \delta(t), \tag{3}$$

$$u|_{x=0} = f(t). \tag{4}$$

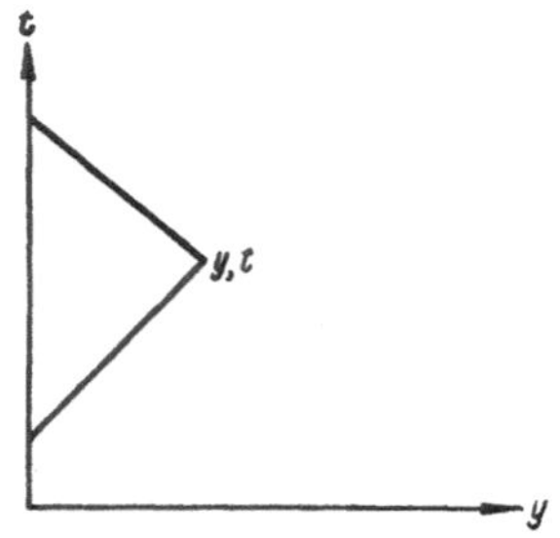

Fig. 1

It is required to find the function $C(x) > 0$. It is assumed that the function $f(t) \in C_1$. The unknown function $C(x)$ is also taken to belong to the class of continuously differentiable functions.

Let us introduce the new independent variable

$$y = \int_0^x \frac{dx}{C(x)}. \tag{5}$$

Then, as a result of simple transformations the problem (1)-(4) can be put into the form

$$u_{tt} = u_{yy} - K(y) u_y, \tag{6}$$

$$u|_{t<0} = 0, \tag{7}$$

$$u_y|_{y=0} = C(0) \delta(t), \tag{8}$$

$$u|_{y=0} = f(t). \tag{9}$$

Here,

$$K(y) \equiv (\ln C(y))'_y. \tag{10}$$

It is not difficult to show that

$$C(0) = -\lim_{t \to +0} f(t) \equiv -f(0), \tag{11}$$

so that condition (8) can be rewritten as

$$u_y|_{y=0} = -f(0) \delta(t). \tag{8a}$$

Let us introduce the notation

$$K(y) u_y(y, t) \equiv G(y, t), \tag{12}$$

and let us solve the problem for Eq. (6) with Cauchy conditions (8), and (9) at $y = 0$, assuming that the function $G(y,t)$ is known:

$$u(y, t) = \frac{1}{2} \iint_{\Delta_{y,t}} G(\eta, \tau) \, d\eta \, d\tau + \frac{1}{2} f(t \pm y). \tag{13}$$

Here, as everywhere in the following, the term containing the sign $\pm$ is to be understood as the sum of two terms corresponding to the choice of the upper and lower signs. The domain of integration in (13) is $\Delta_{y,t}$, a triangle formed by the characteristics of Eq. (6) and the axis $y = 0$ (see Fig. 1). This solution is valid for $t > y$. After a differentiation of (13) with respect to y, we find that

$$u_y(y, t) = \frac{1}{2} \int_0^y G(\eta, t \pm (y - \eta)) \, d\eta \pm \frac{1}{2} f'(t \pm y). \tag{14}$$

Equation (14) contains two unknown functions: $G(y,t)$ and $u_y(y,t)$. Let us derive a second equation for them. It is easy to show that the function u has a discontinuity at $y = t$:

$$u(y,\ y+0)-u(y,\ y-0)=f(0)\exp\left[\frac{1}{2}\int_0^y K(\eta)\,d\eta\right].$$

But, since $u|_{t<y} = 0$ [this follows from the initial condition (7)], we have

$$u(y,\ y+0)=f(0)\ \exp\left[\frac{1}{2}\int_0^y K(\eta)\,d\eta\right]. \tag{15}$$

Substituting (15) into (13), we obtain

$$f(0)\exp\left[\frac{1}{2}\int_0^y K(\eta)\,d\eta\right]=\frac{1}{2}\iint_{\Delta_{y,\,y}} G(\eta,\ \tau)\,d\eta\,d\tau+\frac{1}{2}\,[f(2y)+f(0)]. \tag{16}$$

Differentiating this with respect to y, we obtain

$$\frac{1}{2}K(y)=\frac{\int_0^y G(\eta,2y-\eta)\,d\eta+f'(2y)}{\frac{1}{2}\iint_{\Delta_{y,\,y}} G(\eta,\tau)\,d\eta\,d\tau+\frac{1}{2}\,[f(2y)+f(0)]}. \tag{17}$$

Multiplying (17) and (14) and taking (12) into account, we find that

$$G(y,\ t)=2\ \frac{\left\{\frac{1}{2}\int_0^y G(\eta,t\pm(y-\eta))\,d\eta\pm\frac{1}{2}f'(t\pm y)\right\}\left\{\int_0^y G(\eta,2y-\eta)\,d\eta+f'(2y)\right\}}{\frac{1}{2}\iint_{\Delta_{y,\,y}} G(\eta,\tau)\,d\eta\,d\tau+\frac{1}{2}\,[f(2y)+f(0)]}. \tag{18}$$

Let us introduce the linear operators

$$\left.\begin{aligned}&\Phi G(t,\ y)\equiv\frac{1}{2}\int_0^y G(\eta,\ t\pm(y-\eta))\,d\eta,\\ &\Psi\, G(y)\equiv\int_0^y G(\eta,\ 2y-\eta)\,d\eta,\ \Theta\, G(y)=\frac{1}{2}\iint_{\Delta_{y,\,y}} G(\eta,\ \tau)\,d\eta\,d\tau\end{aligned}\right\} \tag{19}$$

and the abbreviations

$$\left.\begin{aligned}&\pm\frac{1}{2}f'(t\pm y)\equiv\varphi(t,\ y),\quad f'(2y)\equiv\psi(y),\\ &\frac{1}{2}f(2y)+\frac{1}{2}f(0)\equiv-\vartheta(y).\end{aligned}\right\} \tag{20}$$

As a result, Eq. (18) assumes the form

$$G=-2\,\frac{(\Phi G+\varphi)(\Psi G+\psi)}{\vartheta-\Theta G}. \tag{21}$$

Equation (21) is a nonlinear Volterra integral equation of the second kind and, therefore, the theorem on existence and uniqueness holds when the point y, t ($t \geq y$) belongs to the domain $y < y_0$ for a sufficiently small y_0. Let us find y_0, using for this the following variant of the successive approximation method. Let us introduce the parameter λ into Eq. (21):

$$G = -2\frac{(\lambda\Phi G + \varphi)(\lambda\Psi G + \psi)}{\vartheta - \lambda^2\Theta G}, \tag{22}$$

and let us express the unknown function G as a power series in λ:

$$G = G_0 + \lambda G_1 + \lambda^2 G_2 + \ldots \tag{23}$$

Substituting the series (23) into (22), expanding the right-hand side of (22) into powers of λ, and equating the coefficients of equal powers of λ, we obtain

$$G_k = -\frac{2}{\vartheta} \sum_{i+j+\sum_\alpha (\alpha+2) r_\alpha = k} C\{r_\alpha\}\, \Phi G_{j-1} \Psi G_{i-1} \prod_\alpha \left(\frac{1}{\vartheta}\Theta G_\alpha\right)^{r_\alpha}. \tag{24}$$

Here, $C\{r_\alpha\} > 0$ are numerical coefficients depending on the choice of the numbers r_α.

Let us show that all G_k satisfy the estimates

$$|G_k| \leqslant \beta_k y^k. \tag{25}$$

It is obvious that estimate (25) holds for k = 0, provided that $|f'(t)| < A$, $-[f(0) + f(2y)]/2 > \alpha > 0$. It should be noted that if it holds for all G_j with $j<k$, then it also holds for G_k. Indeed, we have

$$|G_k| \leqslant \frac{2}{|\vartheta|} \sum_{i+j+\sum(\alpha+2) r_\alpha = k} C\{r_\alpha\}\, |\Phi G_{j-1}|\, |\Psi G_{i-1}| \prod_\alpha \left(\frac{1}{|\vartheta|} |\Theta G_\alpha|\right)^{r_\alpha} \leqslant$$

$$\leqslant \frac{2}{a} \sum_{i+j+\sum(\alpha+2) r_\alpha = k} C\{r_\alpha\}\, \beta_{j-1} \beta_{i-1} \frac{1}{a^{\sum r_\alpha}} \prod_\alpha \beta_\alpha^{r_\alpha} (\Theta y^\alpha)^{r_\alpha} \Phi y^{j-1} \Psi y^{i-1}.$$

However,

$$\Phi y^{j-1} = \frac{1}{2}\int_0^y 2\eta^{j-1}\, d\eta = \frac{y^j}{j}, \quad \Psi y^{i-1} = \int_0^y \eta^{i-1}\, d\eta = \frac{y^i}{i},$$

and

$$\Theta y^\alpha = \frac{1}{2} \iint_{\Delta_{y,\,y}} \eta^\alpha\, d\eta\, d\tau = \int_0^y (y-\eta)\eta^\alpha\, d\eta = \frac{y^{\alpha+2}}{(\alpha+1)(\alpha+2)}.$$

Therefore, we have

$$|G_k| \leqslant 2 \sum_{i+j+\sum(\alpha+2) r_\alpha = k} C\{r_\alpha\} \frac{\beta_{j-1}\beta_{i-1}}{ji} a^{-1-\sum r_\alpha} \prod_\alpha \left[\frac{\beta_\alpha}{(\alpha+1)(\alpha+2)}\right]^{r_\alpha} y^{\sum(\alpha+2) r_\alpha + i + j} = \beta_k y^k,$$

where

$$\beta_k = 2 \sum_{i+j+\sum(\alpha+2) r_\alpha = k} C\{r_\alpha\} \frac{\beta_{j-1}\beta_{i-1}}{ji} a^{-1-\sum r_\alpha} \prod_\alpha \left[\frac{\beta_\alpha}{(\alpha+1)(\alpha+2)}\right]^{r_\alpha}. \tag{26}$$

Our next step will be to determine the radius of convergence y_0 of the series

$$\sum_{k=0}^{\infty} \beta_k y^k \equiv F(y). \tag{27}$$

We will find the explicit form of the function F(y). To do this, we consider the equation

$$\widetilde{F} = 2\frac{(\Phi\widetilde{F}+A)(\Psi\widetilde{F}+A)}{a-\Theta\widetilde{F}}$$

and we note that if we apply the successive approximation method to this equation we will find that $\widetilde{F}_k(y,t) = \beta_k y^k$ and, thus, $\widetilde{F}(y,t)$ is nothing else but F(y). From this we find that F(y) satisfies the integral equation

$$F(y) = 2\frac{\left(\int_0^y F(\eta)\,d\eta + A\right)^2}{a - \int_0^y (y-\eta)F(\eta)\,d\eta}. \tag{28}$$

Integral equation (28) can be easily solved in explicit form. For this we introduce the function $z(y) \equiv \int_0^y (y-\eta)F(\eta)\,d\eta$. Then, Eq. (28) will transform into the differential equation for z(y)

$$z''(a-z) = 2(z'+A)^2 \tag{29}$$

with the initial conditions

$$z\big|_{y=0} = z'\big|_{y=0} = 0. \tag{30}$$

The solution of (29)-(30) in parametric form is:

$$\left.\begin{aligned} z &= a\left(1 - q\,e^{\frac{1}{2}(1-q^2)}\right), \\ y &= -\frac{a}{A}\int_1^q e^{\frac{1}{2}(1-s^2)} s^2\,ds. \end{aligned}\right\} \tag{31}$$

It is obvious that as the function z(q) is an integral function, the singularities of z(y) are contained among the singularities of q(y). But since y(q) is an integral function, q(y) is analytic with possible branch points given by the roots of the function

$$y'_q = -\frac{a}{A} e^{\frac{1}{2}(1-q^2)} q^2. \tag{32}$$

The only root of the derivative y'_q is situated at q = 0. Thus, the radius of convergence of series (27) is given by

$$y_0 = -\frac{a}{A}\int_1^0 e^{\frac{1}{2}(1-s^2)} s^2\,ds = \frac{a}{A}\int_0^1 e^{\frac{1}{2}(1-s^2)} s^2\,ds \approx 0.410\frac{a}{A}. \tag{33}$$

After Eq. (21) has been solved, it only remains to note that

$$\vartheta(y)-\Theta G=-f(0)\exp\left[\frac{1}{2}\int_0^y K(\eta)\,d\eta\right]=C(0)e^{\frac{1}{2}\ln\frac{C(y)}{C(0)}}=\sqrt{C(0)C(y)}$$

and

$$C(y)=\frac{(\vartheta(y)-\Theta G)^2}{-f(0)}. \tag{34}$$

After the function C(y) has been found, it is not difficult to find C(x). In fact, since $y'_x = 1/C(x) = 1/C(y)$, we have

$$x(y)=\int_0^y C(y)\,dy. \tag{35}$$

Inverting the function (35), we find y(x) and C(x) = C[y(x)].

Note 1. It is obvious from formulas (6) and (15) that the function u(y, t) is the solution of the problem for the differential equation with a displaced argument for t > y

$$u_{tt}=u_{yy}-2\frac{d}{dy}(\ln u(y,\ y+0))\,u_y \tag{36}$$

with the boundary conditions (8) and (9)

$$u|_{y=0}=f(t),\quad u_y|_{y=0,\ t>0}=0.$$

Note 2. With stronger assumptions concerning the smoothness of the function $f(t)$ and the unknown function C(x) we can transform the problem (6)-(9) into the form

$$v_{tt}=v_{yy}+q(y)v,$$
$$v|_{t<0}=0,\quad v|_{y=0}=f(t),\quad v_y|_{y=0}=-f(0)\delta(t)-\frac{f'(0)}{f(0)}f(t).$$

Here, we have

$$v=u\exp\left[-\frac{1}{2}\int_0^y K(\eta)\,d\eta\right],\quad q(y)=\frac{1}{2}K_y-\frac{1}{4}K^2.$$

As can be easily shown, the function $S \equiv qv_t$ satisfies the integral equation

$$-f(0)S(y,\ t)=2\left[\int_0^y S(\eta,\ 2y-\eta)\,d\eta-f''(2y)+\frac{f'(0)}{f(0)}f'(2y)\right]\times$$
$$\times\left[-\frac{1}{2}\iint\limits_{\Delta_{y,t}} S(\eta,\ \tau)\,d\eta\,d\tau+\frac{1}{2}f'(t\pm y)\mp\frac{f'(0)}{2f(0)}f(t\pm y)\right].$$

If we introduce the function

$$K(y,\ t)=\frac{1}{f(0)}\left[-\frac{1}{2}\iint\limits_{\Delta_{y,t}} S(\eta,\ \tau)\,d\eta\,d\tau+\frac{1}{2}f'(t\pm y)\mp\frac{f'(0)}{2f(0)}f(t\pm y)\right],$$

then for K it is easy to obtain the differential equation

$$K_{tt} - K_{yy} = 2\frac{d}{dy}(K(y, y))K(y, t)$$

and the boundary conditions

$$K_y|_{y=0} = -\frac{f'(0)}{f(0)}K|_{y=0}, \quad K|_{y=0} = \frac{f'(t)}{f(0)}.$$

It can be seen from these equations that K(y,t) is the kernel of the Volterra operator transforming the operator $L\varphi = (\partial^2/\partial y^2)\varphi + q(y)\varphi$ with boundary conditions $\varphi'(0) + \gamma\varphi(0) = 0$ $[\gamma = f'(0)/f(0)]$ to the operator $L_0 g = (\partial^2/\partial t^2)g$ with boundary conditions $g'_t|_{t=0} = 0$.

§ 2. The Inverse Seismic Problem

Let us consider the three-dimensional Lamb problem on the propagation of waves in a laminarly inhomogeneous semi-infinite medium $x_1 > 0$. As is known, the displacement vector u in this case satisfies the system of equations

$$(\lambda+\mu)\frac{\partial}{\partial x_i}\operatorname{div} u + \mu\Delta u_i + \frac{\partial\mu}{\partial x_1}\left(\frac{\partial u_1}{\partial x_i} + \frac{\partial u_i}{\partial x_1}\right) + \delta_{i1}\frac{\partial\lambda}{\partial x_1}\operatorname{div} u = \rho\frac{\partial^2 u_i}{\partial t^2}. \tag{37}$$

Here δ_{i1} is the Kronecker symbol.

It is assumed that λ, μ, ρ depend only on x_1. We will assume that the positive functions λ, μ, ρ are unknown, but in addition to the initial conditions

$$u|_{t<0} = 0 \tag{38}$$

and the usual boundary conditions

$$\left.\begin{aligned} \lambda\operatorname{div} u + 2\mu\frac{\partial u_1}{\partial x_1}\Big|_{x_1=0} &= \delta(x_2, x_3)\delta(t), \\ \mu\left(\frac{\partial u_i}{\partial x_1} + \frac{\partial u_1}{\partial x_i}\right)\Big|_{x_1=0} &= \delta(x_2, x_3)\delta(t), \quad i = 2, 3, \end{aligned}\right\} \tag{39}$$

we are also given certain characteristic functions $f_i(t, x_2, x_3)$

$$u_i|_{x_1=0} = f_i(t, x_2, x_3). \tag{40}$$

Let us first of all notice that in view of the fact that system (37) is hyperbolic, the functions $u_i(t, x_1, x_2, x_3)$ are finite for any fixed t and, therefore, moments of any order of the function u_i exist. Let us consider the functions

$$\left.\begin{aligned} U_i(x_1, t) &= \iint u_i\, dx_2\, dx_3, \quad i = 1, 2, \\ V_1(x_1, t) &= \iint u_1 x_2\, dx_2\, dx_3 \end{aligned}\right\} \tag{41}$$

and let us assume that instead of conditions (40) we are given the values of the functions U_1, U_2, and V_1 at $x_1 = 0$,

$$\left.\begin{aligned} U_i|_{x_1=0} &= \iint f_i(t, x_2, x_3)\, dx_2 dx_3 \equiv F_i(t), \quad i = 1, 2, \\ V_1|_{x_1=0} &= \iint f_1(t, x_2, x_3)\, x_2 dx_2 dx_3 \equiv G_1(t). \end{aligned}\right\} \tag{42}$$

The differential equation for U_1 is

$$\rho\frac{\partial^2 U_1}{\partial t^2}=\iint\rho(x_1)\frac{\partial^2 u_1}{\partial t^2}dx_2dx_3=(\lambda+\mu)\iint\frac{\partial}{\partial x_1}\operatorname{div}u\,dx_2dx_3+$$

$$+\mu\iint\Delta u_1dx_2dx_3+2\frac{\partial\mu}{\partial x_1}\iint\frac{\partial u_1}{\partial x_1}dx_2dx_3+\frac{\partial\lambda}{\partial x_1}\iint\operatorname{div}u\,dx_2dx_3.$$

In view of the finiteness of the function u_i, all terms on the right-hand side of the last equation containing derivatives with respect to x_2 and x_3 are zero and, therefore, we have

$$\rho\frac{\partial^2 U_1}{\partial t^2}=(\lambda+2\mu)\frac{\partial^2}{\partial x_1^2}\iint u_1dx_2dx_3+\frac{\partial}{\partial x_1}(\lambda+2\mu)\frac{\partial}{\partial x_1}\iint u_1dx_2dx_3,$$

or

$$\rho\frac{\partial^2 U_1}{\partial t^2}=\frac{\partial}{\partial x_1}\left[(\lambda+2\mu)\frac{\partial}{\partial x_1}U_1\right]. \tag{43}$$

Similarly, we find that the equations for U_2 and V_1 are

$$\rho\frac{\partial^2 U_2}{\partial t^2}=\frac{\partial}{\partial x_1}\left(\mu\frac{\partial}{\partial x_1}U_2\right), \tag{44}$$

$$\rho\frac{\partial^2 V_1}{\partial t^2}=\frac{\partial}{\partial x_1}\left[(\lambda+2\mu)\frac{\partial V_1}{\partial x_1}\right]-(\lambda+\mu)\frac{\partial U_2}{\partial x_1}-\frac{\partial\lambda}{\partial x_1}U_2. \tag{45}$$

Further, instead of (39), the boundary conditions for U_1, U_2, and V_1 are

$$\left.\begin{aligned}&(\lambda+2\mu)\frac{\partial U_1}{\partial x_1}\Big|_{x_1=0}=\delta(t),\\&\mu\frac{\partial U_2}{\partial x_1}\Big|_{x_1=0}=\delta(t),\\&(\lambda+2\mu)\frac{\partial V_1}{\partial x_1}\Big|_{x_1=0}=\lambda(0)F_2(t).\end{aligned}\right\} \tag{46}$$

Let us make the following change of variables in Eqs. (44) and (45), as well as in the corresponding boundary conditions:

$$y_1=\int_0^{x_1}\sqrt{\frac{\rho}{\lambda+2\mu}}\,dx_1,\quad y_2=\int_0^{x_1}\sqrt{\frac{\rho}{\mu}}\,dx_1. \tag{47}$$

We now have for U_i

$$U_{itt}=U_{iy_iy_i}-K_i(y_i)U_{iy_i}, \tag{48}$$

$$U_i|_{y_i=0}=F_i(t), \tag{49}$$

$$U_{iy_i}|_{y_i=0}=a_i(0)\delta(t), \tag{50}$$

$$U_i|_{t<0}=0. \tag{51}$$

Here,

$$K_i(y_i)\equiv\frac{\partial}{\partial y_i}(\ln a_i(y_i)), \tag{52}$$

$$a_1(y_1) \equiv \frac{1}{\sqrt{(\lambda+2\mu)\rho}}, \quad a_2(y_2) = \frac{1}{\sqrt{\mu\rho}}. \tag{53}$$

It should be noted that the problems (48)-(51) are identical with problem (6)-(9) and, therefore, the functions $a_i(y_i)$ and $U_i(y_i, t)$ are found by the method described in § 1 or by any other method.

It now remains for us to solve two closely related problems:

1. to find the functions λ, μ, and ρ separately;
2. to transform the coordinates y_1, y_2 to the variable x_1.

Let us note first of all that it is not difficult to establish the connection between the coordinates y_1 and y_2. Indeed, it follows from (47) that

$$\sqrt{\rho(\lambda+2\mu)}\,dy_1 = \sqrt{\rho\mu}\,dy_2 = \rho\,dx_1,$$

or

$$\frac{dy_1}{a_1(y_1)} = \frac{dy_2}{a_2(y_2)}. \tag{54}$$

Integrating (54), we can easily find y_2 as a function of y_1. Transforming now to the variable y_1 in Eq. (45), we obtain

$$V_{1tt} = V_{1y_1y_1} - K_1(y_1)V_{1y_1} - (\lambda+\mu)a_1(y_1)\frac{\partial U_2(y_1, t)}{\partial y_1} - a_1\frac{\partial\lambda}{\partial y_1}U_2. \tag{55}$$

In view of what has been said above, the functions K_1, a_1, a_2, and U_2 can be considered to be known (including U_2 and a_2 as functions of y_1). Equations (43) and (44) describe the propagation of waves with the velocities of the longitudinal and transverse waves, respectively. Since, with the variables y_1, t the velocity of the longitudinal waves is unity, the velocity of the transverse waves is less than $1/\sqrt{2}$ [we will not discuss here the conditions that must be imposed on $F_i(t)$ for this to hold]. Therefore, $U_2(y_1, t) \equiv 0$ for $t < y_1$ and, therefore, we also have $V_1\big|_{t<y_1} \equiv 0$. Since V_1 satisfies the boundary conditions

$$V_1\big|_{y_1=0} = G_1(t), \quad \frac{\partial V_1}{\partial y_1}\bigg|_{y_1=0} = a_1(0)\lambda(0)F_2(t), \tag{56}$$

then

$$V_1(y_1, y_1) = G_1(0)\exp\left[\frac{1}{2}\int_0^{y_1} K_1(\eta_1)\,d\eta_1\right]. \tag{57}$$

Our most immediate problem now is to derive the system of integral equations for the functions $w = V_{1y_1}$ and λ. For this, we write down the solutions of the problem for the equation $V_{1tt} - V_{1y_1y_1} = H(y_1, t)$ with Cauchy boundary conditions (56) at $y_1 = 0$. Substituting $H(y_1, t)$ from (55), we obtain

$$V_1(y_1, t) = \frac{1}{2}\iint_{\Delta_{y_1t}} K_1(\eta_1)V_{1\eta_1}(\eta_1, \tau)\,d\eta_1\,d\tau + \frac{1}{2}\iint_{\Delta_{y_1t}} (\lambda(\eta_1) +$$

$$+ \mu(\eta_1))a_1(\eta_1)\frac{\partial U_2(\eta_1, \tau)}{\partial\eta_1}\,d\eta_1\,d\tau + \frac{1}{2}\iint_{\Delta_{y_1t}} a_1(\eta_1)\frac{\partial\lambda(\eta_1)}{\partial\eta_1}U_2(\eta_1, \tau)\,d\eta_1\,d\tau + \tag{58}$$

$$+\frac{1}{2}G_1(t\pm y_1)+\frac{1}{2}a_1(0)\lambda(0)\int\limits_{t-y_1}^{t+y_1}F_2(\tau)\,d\tau=\frac{1}{2}\iint\limits_{\Delta_{y_1t}}K_1(\eta_1)\times$$

$$\times V_{1\eta_1}(\eta_1,\tau)\,d\eta_1\,d\tau+\frac{1}{2}\iint\limits_{\Delta_{y_1t}}\mu(\eta_1)a_1(\eta_1)\frac{\partial U_2}{\partial\eta_1}\,d\eta_1\,d\tau-$$

$$-\frac{1}{2}\iint\limits_{\Delta_{y_1t}}\frac{\partial a_1}{\partial\eta_1}\lambda U_2\,d\eta_1\,d\tau+\frac{1}{2}\int\limits_0^{y_1}a_1(\eta_1)\lambda(\eta_1)U_2(\eta_1,\ t\pm(y_1-\eta_1))\,d\eta_1+$$

$$+\frac{1}{2}G_1(t\pm y_1). \tag{58}$$

It is obvious that all terms on the right-hand side of (58) are automatically continuous on the line $t = -y_1$, with the possible exception of $\frac{1}{2}G(y_1 + t)$. However, since $V(y_1, t) \equiv 0$ when $t < 0$, then $G(0) = 0$ (necessary condition for solvability) and, therefore, in view of (57) we have

$$V_1(y_1,\ y_1)\equiv 0. \tag{59}$$

Let us write

$$\mu=\sigma\lambda, \tag{60}$$

where $\sigma = \mu/\lambda = \{[(\lambda + 2\mu)/\mu] - 2\}^{-1} = a_1^2/(a_2^2 - 2a_1^2)$ is a known function.

Differentiating (58) with respect to y_1, we obtain

$$w(y_1,\ t)=\frac{1}{2}\int\limits_0^{y_1}K_1(\eta_1)w(\eta_1,\ t\pm(y_1-\eta_1))\,d\eta_1+\frac{1}{2}\int\limits_0^{y_1}\sigma(\eta_1)a_1(\eta_1)\times$$

$$\times\lambda(\eta_1)\frac{\partial U_2(\eta_1,\ t\pm(y_1-\eta_1))}{\partial\eta_1}\,d\eta_1-\frac{1}{2}\int\limits_0^{y_1}\frac{\partial a_1(\eta_1)}{\partial\eta_1}\lambda(\eta_1)U_2(\eta_1,\ t\pm$$

$$\pm(y_1-\eta_1))\,d\eta_1+a_1(y_1)\lambda(y_1)U_2(y_1,\ t)\pm\frac{1}{2}\int\limits_0^{y_1}a_1(\eta_1)\lambda(\eta_1)\times$$

$$\times\frac{\partial U_2(\eta_1,\ t\pm(y_1-\eta_1))}{\partial\tau}\,d\eta_1\pm\frac{1}{2}G_1'(t\pm y_1).$$

Here, $(\partial U_2/\partial\eta_1)\,[\eta_1,\ t\pm(y_1-\eta_1)]$ is taken to mean the derivative of U_2 with respect to the first argument, the fact that the derivative of the discontinuity is a δ function being taken into account. The function $U_2(y_1, t)$ has a discontinuity on the line $t=t(y_1)\equiv\int\limits_0^{y_1}\frac{a_1(\eta_1)}{a_2(\eta_1)}\,d\eta_1$, corresponding to the wave front of the transverse wave. We thus have

$$U_2|_{t<t(y_1)}=0,\ U_2|_{t=t(y_1)+0}=F_2(0)\exp\left[\frac{1}{2}\int\limits_0^{y_1}K_2(\eta_2)\,d\eta_2\right]=-\sqrt{a_2(y_1)a_2(0)}.$$

Therefore,

$$w(y_1, t) = \frac{1}{2}\int_0^{y_1} K_1(\eta_1)\, w(\eta_1, t \pm (y_1 - \eta_1))\, d\eta_1 + \frac{1}{2}\int_0^{y_1} \sigma(\eta_1)\, a_1(\eta_1) \times$$

$$\times \lambda(\eta_1) \frac{\partial' U_2(\eta_1, t \pm (y_1 - \eta_1))}{\partial' \eta_1}\, d\eta_1 - \frac{1}{2}\int_0^{y_1} \frac{\partial a_1(\eta_1)}{\partial \eta_1} \lambda(\eta_1)\, U_2(\eta_1, t \pm$$

$$\pm (y_1 - \eta_1))\, d\eta_1 \pm \frac{1}{2}\int_0^{y_1} a_1(\eta_1)\lambda(\eta_1) \frac{\partial'}{\partial' \tau} U_2(\eta_1, t \pm (y_1 - \eta_1))\, d\eta_1 +$$

$$+ a_1(y_1)\lambda(y_1)\, U_2(y_1, t) + \frac{1}{2}\varepsilon(t(y_1) - t)\left[\frac{\sigma(y_1^{\pm})\, a_1^2(y_1^{\pm})\lambda(y_1^{\pm})}{a_2(y_1^{\pm}) \pm a_1(y_1^{\pm})} \times\right.$$

$$\left.\times \sqrt{a_2(y_1^{\pm})\, a_2(0)} \mp \frac{a_1(y_1^{\pm})\lambda(y_1^{\pm})\, a_2(y_1^{\pm})}{a_2(y_1^{\pm}) \pm a_1(y_1^{\pm})} \sqrt{a_2(y_1^{\pm})\, a_2(0)}\right] \pm \frac{1}{2} G_1'(t \pm y_1). \tag{61}$$

Here, $y_1^{\pm}$ are the roots of the equations

$$t - t(y_1^+) + (y_1 - y_1^+) = 0, \quad t - t(y_1^-) - (y_1 - y_1^-) = 0.$$

The symbol $\partial'/\partial'\eta_1$ denotes that U_2 is to be differentiated as a classical function only in the intervals where it is continuous.

We obtain the second integral equation by the substitution of (58) into (59),

$$V_1(y_1, y_1) = \frac{1}{2}\iint_{\Delta_{y_1 y_1}} K_1(\eta_1)\, w(\eta_1, \tau)\, d\eta_1\, d\tau + \frac{1}{2}\iint_{\Delta_{y_1 y_1}} \sigma(\eta_1)\, a_1(\eta_1) \times$$

$$\times \lambda(\eta_1) \frac{\partial U_2}{\partial \eta_1}\, d\eta_1\, d\tau - \frac{1}{2}\int_0^{y_1} a_1(\eta_1)\lambda(\eta_1)\, U_2(\eta_1, 2y_1 - \eta_1)\, d\eta_1 +$$

$$+ \frac{1}{2}\int_0^{y_1} a_1(\eta_1)\lambda(\eta_1)\, U_2(\eta_1, 2y_1 - \eta_1)\, d\eta_1 + \frac{1}{2} G(2y_1) = 0. \tag{62}$$

In (62) we have taken into account that $U_2(\eta_1, \eta_1) \equiv 0$. After differentiating (62) with respect to y_1, and taking into account that the derivatives of U_2 contain δ functions, we find that

$$\int_0^{y_1} K_1(\eta_1)\, w(\eta_1, 2y_1 - \eta_1)\, d\eta_1 + \int_0^{y_1} \sigma(\eta_1)\, a_1(\eta_1)\lambda(\eta_1) \times$$

$$\times \frac{\partial' U_2(\eta_1, 2y_1 - \eta_1)}{\partial' \eta_1}\, d\eta_1 - \int_0^{y_1} \frac{\partial a_1(\eta_1)}{\partial \eta_1} \lambda(\eta_1)\, U_2(\eta_1, 2y_1 - \eta_1)\, d\eta_1 +$$

$$+ \int_0^{y_1} a_1(\eta_1)\lambda(\eta_1) \frac{\partial' U_2(\eta_1, 2y_1 - \eta_1)}{\partial' \tau}\, d\eta_1 + \sigma(y_1^0)\, a_1^2(y_1^0)\lambda(y_1^0) \times$$

$$\times \frac{\sqrt{a_2(y_1^0)\, a_2(0)}}{a_2(y_1^0) + a_1(y_1^0)} - a_1(y_1^0)\lambda(y_1^0)\, a_2(y_1^0) \frac{\sqrt{a_2(y_1^0)\, a_2(0)}}{a_2(y_1^0) + a_1(y_1^0)} + G'(2y_1) = 0. \tag{63}$$

Here y_1^0 is the root of the equation $2y_1 - t(y_1^0) - y_1^0 = 0$. The coefficient of $\lambda(y_1^0)$ in (63) can be written as

$$\varkappa(y_1^0) = \frac{a_1(y_1^0)\sqrt{a_2(y_1^0)\,a_2(0)}}{a_2(y_1^0)+a_1(y_1^0)}\left[a_1(y_1^0)\,\sigma(y_1^0) - a_2(y_1^0)\right].$$

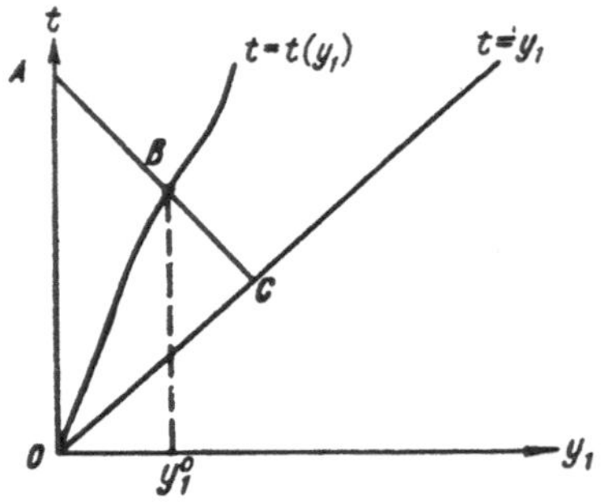

Fig. 2.

In the following we will require that $a_1(y_1^0)\sigma(y_1^0) - a_2(y_1^0)$ be nonzero in a certain interval $[0, \bar{y}_1]$. Since σ, a_1, and a_2 are interconnected, this requirement means that the quantity $\sigma(y_1^0)$ nowhere coincides with the positive root of the equation $\sigma^3 - 2\sigma - 1 = 0$, i.e., with $\sigma_0 = (1+\sqrt{5})/2 \approx 1.618$. If the last condition is satisfied, then Eq. (63) can be divided by $\varkappa(y_1^0)$; then (61) and (63) form a system of integral equations for λ and w of the form

$$\left.\begin{aligned} w(y_1, t) &= Kw + L\lambda + M\lambda + g_1, \\ \lambda(y_1) &= Pw + Q\lambda + g_2. \end{aligned}\right\} \qquad (64)$$

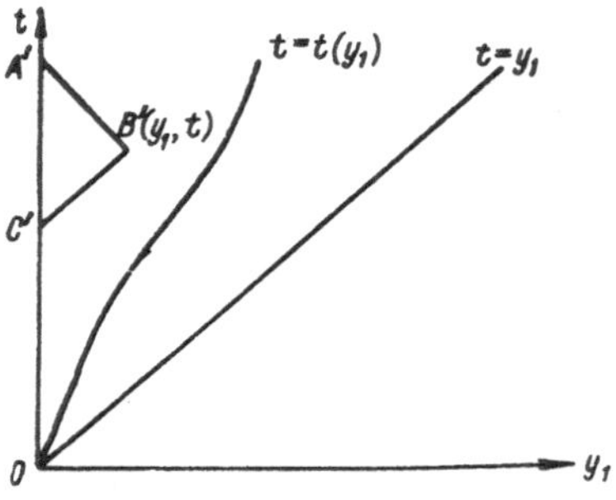

Fig. 3.

The structure of the operators K, L, P, Q, and M is the following:

P is the operator of integration over AC;

Q is the operator of integration over AB (see Fig. 2);

K is the operator of integration over A'B' and B'C' (see Fig. 3).

For $t > t(y_1)$:

L is the operator of integration over A'B' and B'C';

M is the operator of multiplication by a bounded function; we denote it by $M_{B'}$, and we will assume that $M_{B'} = 0$ for $t < t(y_1)$ (see Fig. 3).

For $y_1 < t < t(y_1)$:

L is the operator of integration over A"D and C"E;

M is the operator of multiplication by a bounded function at the points D and E (Fig. 4). We adopt the notation $M = M_D + M_E + M_{B'}$, assuming that $M_D = M_E = 0$ when $t > t(y_1)$. All integration operators have bounded kernels.

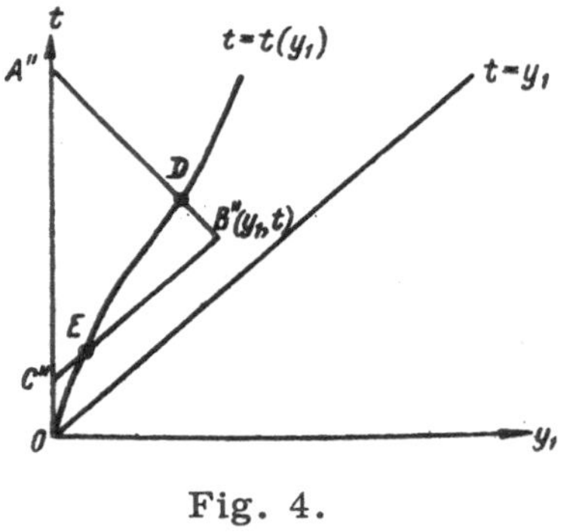

Fig. 4.

Since K is a volterra operator, the operator E − K can be inverted, and then

$$w = (E-K)^{-1}L\lambda + (E-K)^{-1}M\lambda + g_3,$$

where $g_3 \equiv (E-K)^{-1}g_1$; $(E-K)^{-1}L$ is obviously a Volterra operator; $(E-K)^{-1}M = M + RM$.

The operator R is an integral one, but the operator RM can be decomposed into a sum of an integral operator and an operator of the type of multiplication by a bounded function. Indeed, the operator R is a sum of the operators of integration over A'B', C'B', and the interior of the triangle A'B'C' (see Fig. 3). It is easy to see that $RM = R_{A'B'}M_D + R_{C'B'}M_E + N$, where N is the Volterra operator. The term $R_{A'B'}M_D$ can be easily calculated explicitly. It is the operator of multiplication of the value of the function at the point D by the function $\frac{1}{2}\varkappa(y_1^0) \times$ $\times \left\{\exp\left[\int\limits_{y_1^0}^{y_1} K_1(\eta_1)\,d\eta_1\right] - 1\right\}$. The solution of Eq. (64) for w can be written as

$$w = S\lambda + M_1\lambda + g_3, \tag{65}$$

where S is a Volterra operator and M_1 has the same structure as M.

Substituting (65) into the second of equations (64), we find that

$$\lambda(y_1^0) = PS\lambda + PM_1\lambda + Q\lambda + g_4.$$

The operators PS and Q are obviously Volterra operators, while the operator PM_1 is the sum of a Volterra operator and the operator $PM_{1D} = P_{BC}M_{1D}$. As can be easily established, the last operator is the operator of multiplication at the point D by the function

$$1 - \exp\left[\frac{1}{2}\int\limits_{y_1^0}^{\hat{y}_1} K_1(\eta_1)\,d\eta_1\right] = 1 - \sqrt{\frac{a_1(\hat{y}_1)}{a_1(y_1^0)}},$$

where $\hat{y}_1 = [t(y_1^0) + y_1^0]/2$. The second of equations (64) can now be written as

$$\lambda(y_1^0) = T\lambda + g_4, \tag{66}$$

where T is a Volterra operator and, therefore, the solution of Eq. (66) exists and is unique.

After the function $\lambda(y_1)$ has been determined, $\mu(y_1)$ and $\rho(y_1)$ can be easily found from formulas (53), while the first of relations (47) allows us, as in § 1, to obtain these functions as functions of the variable x_1.

Literature Cited

1. M.G Krein, The transition function of the one-dimensional second-order boundary problem, Dokl. Akad. Nauk SSSR, 88(3) : 405-408 (1953).
2. M.G. Krein, A method for the effective solution of the inverse problem, Dokl. Akad. Nauk SSSR, 94(6) : 987-990 (1954).
3. I.M. Gel'fand and B.M. Levitan, On the definition of a differential equation by its spectral functions, Izv. Akad. Nauk SSSR, seriya matem., 15(4) : 309-360 (1951).
4. A.S. Alekseev, Some inverse problems in the theory of wave propagation, Parts I and II, Izv. Akad. Nauk SSSR, seriya geofiz., 15(11) : 1514-1531 (1962).

THE TRACE FORMULAS AND SOME ASYMPTOTIC ESTIMATES OF THE RESOLVENT KERNEL OF THE THREE-DIMENSIONAL SCHROEDINGER EQUATION

V. S. Buslaev

Introduction

The present paper is devoted to a detailed description of the results summarized in the author's preceding note [1]. The aim of [1] was the generalization to the three-dimensional case of the trace formulas obtained earlier for the one-dimensional Schroedinger operator with a potential decreasing at infinity [2].

Trace formulas are identities for the spectral characteristics of an operator and they can be interpreted as explicit expressions for the spectral traces of integral powers of the operator regularized in a definite sense. Such identities were initially introduced by Gel'fand and Levitan [3] for the case of a regular one-dimensional Schroedinger operator (for a review of the subject, see [4]*), and then some of the formulas of this type were also found for an operator with a continuous spectrum — the Schroedinger operator for a half-space with a decreasing potential (see Levinson [7], Newton [8], and Faddeeva [9]). A rigorous derivation of the complete series of trace formulas for the latter problem has been given in [2].† In [1], analogous identities were obtained for the three-dimensional Schroedinger operator.

The characteristics of the operator spectrum contained in the trace formulas are regularized moments of Krein's spectral shift function [13]. It has been found that this function in the concrete examples of problems with a continuous spectrum can be expressed in terms of the determinant of the scattering matrix.‡ An elegant formula which has given this result a general significance was proved by Birman and Krein [15] within the framework of the abstract approach to scattering theory.

*See also two papers by Lomonosov [5, 6] which were not covered in this review [4].

† Percival [10] has given another method for the derivation of these formulas and subsequently used it with Roberts [11] in the case of the radial Schroedinger operator with centrifugal terms. Another paper by Roberts [12] should also be noted in which trace formulas are used for numberical estimates of the limiting scattering phase.

‡ See also Lomonosov's paper [14] in which a related result was obtained independently for the three-dimensional case.

Let us also note a recent note by Berezin [16] in which relations of the trace-formula type are derived for the many-particle Schroedinger operator in the quantum statistical mechanics.

In the present paper, in addition to a detailed presentation of the results of [1], we will also give a description of some asymptotic estimates of the resolvent kernel and of the solutions of the scattering problem. In § 1 we consider the behavior of the kernel of the resolvent of the Schroedinger equation and the solutions of the scattering problem at large distances and large values of the spectral parameter. In § 2 we derive a formula connecting the spectral shift function with the scattering operator. This formula is now a particular case of Birman and Krein's result [15], although our derivation of it is essentially different from that given in [15]. In § 3 we show how the results of the preceding sections are used to obtain trace formulas.

§ 1. Asymptotic Estimates

Let H be a self-adjoint operator in $L_2(E_3)$ defined by the differential expression

$$-\Delta u(x)+q(x)\,u(x)\quad (x\in E_3). \tag{1.1}$$

Let us assume that q(x) is a real function possessing derivatives of any order and that it decreases together with its derivatives faster than any power of $|x|^{-1}$ as $|x|\to\infty$. Moreover, we will assume that the equation

$$-\Delta u(x)+q(x)\,u(x)=0 \tag{1.2}$$

does not have any nonzero solutions u(x) satisfying the condition

$$\left(\frac{x}{|x|},\ \nabla u(x)\right)_{|x|\to\infty}=O\left(\frac{1}{|x|^2}\right).$$

Let R_λ denote the resolvent of the operator H. A superscript "zero" will be used to denote quantities evaluated at q = 0, for example, $R_\lambda^0 = R_\lambda \big|_{q=0}$.

Everywhere in the present paper we will use, without special references, the results of Povzner [17, 18], Kato [19], and Ikebe [20] concerning the properties of the resolvent R_λ of the operator H and the spectral properties of the operator H, as well as numerous results on the scattering problem quoted in these papers. Assertions which can be easily obtained on the basis of [14-16] will also not be specifically distinguished.

The spectrum of the operator consists of a continuous spectrum over $[0, \infty)$ and a finite number of eigenvalues λ_l ($\lambda_l \le 0$, $l = 1, 2, \ldots, M$) of finite multiplicity m_l. It is not difficult to show that on account of the assumption made above concerning the solutions of Eq. (1.2), the point $\lambda = 0$ cannot be an eigenvalue of the operator H. The resolvent R_λ ($\operatorname{Im}\lambda \ne 0$) is an integral operator in $L_2(E_3)$. The difference

$$R_\lambda(x,\ x')-R_\lambda^0(x,\ x')=R_\lambda(x,\ x')-\frac{e^{i\sqrt{\lambda}|x-x'|}}{4\pi|x-x'|}$$

depends analytically on the variable λ with the exception of the spectrum points; $\sqrt{\lambda}$ is defined in the plane λ with a cut $\lambda \ge 0$ such that $\operatorname{Im}\sqrt{\lambda} \ge 0$.

If λ tends to the points on the half-axis $[0, \infty)$, always remaining in the upper or lower half-plane, then the function

$$R_\lambda(x,\ x')-R_\lambda^0(x,\ x')$$

assumes various limiting values uniformly with respect to x and x' of the compact subdomain $E_3 \times E_3$. These limiting values are infinitely differentiable with respect to $\sqrt{\lambda}$ for $\lambda \geq 0$.

Several of the following propositions describe the behavior of $R_{\lambda+i0}(x,x')$ $(\lambda > 0)$ for large $|x|$ and $|x'|$. The asymptotic behavior of $R_\lambda(x,x')$ as $|\lambda| \to \infty$ is described in the second part of this section.

Most of the proofs in this section are carried out as follows. Usually, there exists a simple method (integration by parts of the expansion of a regular function into a power series) which allows us to describe easily the nature of the asymptotic expression, to estimate its error, and to establish the possibility of its differentiability. However, an explicit form for the coefficients of the asymptotic expression can only be obtained with difficulty in this way. We will obtain such expressions by substituting the asymptotic formula into the differential equation and taking its differentiability into account.

Theorem 1

The following asymptotic formula is valid as $x \to \infty$:

$$R_{\lambda+i0}(x,\ x') = R^0_{\lambda+i0}(x,\ x') +$$
$$+ R^0_{\lambda+i0}(x,\ 0)\sum_{n=0}^{N}\frac{1}{n!}\left(\frac{1}{2i\sqrt{\lambda}|x|}\right)^n B_n(\Delta_\alpha)\left[\psi_{-\alpha}(x',\ \lambda) - \psi^0_{-\alpha}(x',\ \lambda)\right] +$$
$$+ Q^N_\lambda(x,\ x') \sim R^0_{\lambda+i0}(x,\ 0)\sum_{n=0}^{\infty}\frac{1}{n!}\left(\frac{1}{2i\sqrt{\lambda}|x|}\right)^n B_n(\Delta_\alpha)\psi_{-\alpha}(x',\ \lambda). \tag{1.3}$$

In this formula, $\lambda > 0$, $\sqrt{\lambda} \equiv \sqrt{\lambda + i0}$, $\alpha = x/|x|$; Δ_α is the spherical Laplace operator acting on the function depending on the unit vector α

$$\Delta_\alpha f(\alpha) = |x|^2\,\Delta f\left(\frac{x}{|x|}\right);$$

$B_n(x)$ are n-th order polynomials of x: $B_0(x) \equiv 1$ and $B_{n+1}(x) = [x + n(n+1)]B_n(x)$; $\psi_\alpha(x,\ \lambda)$ is the solution of the so-called scattering problem

$$\psi_\alpha(x,\ \lambda) \equiv e^{i\sqrt{\lambda}x\alpha} - \int dy R_{\lambda+i0}(x,\ y)\,q(y)\,e^{i\sqrt{\lambda}y\alpha}. \tag{1.4}$$

The quantity $Q^N_\lambda(x',x')$ has the estimate

$$|Q^N_\lambda(x,\ x')| \leqslant \frac{C_N}{|x|}\cdot\frac{(1+\sqrt{\lambda})^{N+1}}{|x|^{N+1}}\cdot\frac{1}{1+|x'|}. \tag{1.5}$$

If $q(x) = 0$ for $|x| \geq R$, then the asymptotic series (1.3) becomes convergent for $|x| > R$ and $|x| > |x'|$ and the sign of asymptotic equality in formula (1.3) can be replaced by the equality sign.

Before proving this theorem we will establish the following lemma.

Lemma 1

In the notation of Theorem 1, the following formula holds:

$$R^0_{\lambda+i0}(x, x') =$$
$$= R^0_{\lambda+i0}(x, 0)\sum_{n=0}^{N}\frac{1}{n!}\left(\frac{1}{2i\sqrt{\lambda}|x|}\right)^n B_n(\Delta_\alpha)\psi^0_{-\alpha}(x', \lambda) + R^N_\lambda(x, x'), \tag{1.6}$$

where

$$|R^N_\lambda(x, x')| \leqslant \frac{C_N}{|x|}(1+\sqrt{\lambda}|x'|)^{N+1}\left(\frac{|x'|}{|x|}\right)^{N+1} \tag{1.7}$$

for $|x'|/|x| < 1$, $|x'|^2/|x| \le 1$, and

$$|R^N_\lambda(x, x')| \leqslant \frac{C(\lambda)}{|x|}\left(\frac{|x'|}{|x|}\right)^{N+1} \tag{1.8}$$

for $|x'| \le R$, $R/|x| \le \delta < 1$ and N larger than a certain value $\tilde{N} = \tilde{N}(R, \delta)$.

Proof of the Lemma. We have

$$R^0_{\lambda+i0}(x, x') = \frac{e^{i\sqrt{\lambda}|x-x'|}}{4\pi|x-x'|} = R^0_{\lambda+i0}(x, 0)\, e^{i\sqrt{\lambda}|x'|F_1\left(\frac{|x'|}{|x|}\right)} F_2\left(\frac{|x'|}{|x|}\right).$$

$F_1(z)$ and $F_2(z)$ are regular functions of z for $|z| < 1$. Therefore, with $|x'|/|x| < 1$ the product of the second and third factors can be expanded in inverse powers of $|x|$:

$$R^0_{\lambda+i0}(x, x') = R^0_{\lambda+i0}(x, 0)\sum_{n=0}^{\infty}\frac{1}{n!}\left(\frac{1}{2i\sqrt{\lambda}|x|}\right)^n \rho^{(n)}_\alpha(x', \lambda).$$

Using simple explicit expressions for F_1 and F_2, we easily obtain estimates (1.7) and (1.8) for the remainder term of such an expansion up to the term in $|x|^{n-1}$ ($n \le N$). It is also obvious that the expansion obtained can be differentiated with respect to the coordinates of the point x. Substituting it into the equation

$$(-\Delta_x - \lambda)R^0_{\lambda+i0}(x, x') = \delta(x - x'),$$

we find

$$\rho^{(n+1)}_\alpha(x', \lambda) = [\Delta_\alpha + n(n+1)]\,\rho^{(n)}_\alpha(x', \lambda).$$

The coefficient $\rho^{(0)}_\alpha(x', \lambda)$ can be easily calculated to be

$$\rho^{(0)}_\alpha(x', \lambda) = \psi^0_{-\alpha}(x', \lambda).$$

This ends the proof of the lemma.

Proof of Theorem 1. The function $R_\lambda(x, x')$ satisfies the equation

$$R_\lambda(x, x') = R^0_\lambda(x, x') - \int dy R^0_\lambda(x, y)\, q(y)\, R_\lambda(y, x'). \tag{1.9}$$

It is also known that for $\lambda \ge 0$ we have

$$|R_\lambda(x, x')| \leqslant \frac{C(q)}{|x-x'|}. \tag{1.10}$$

The assertions of Theorem 1 are easily derived from this with the help of Lemma 1.

Theorem 2.1

In the notation of Theorem 1, the following asymptotic expansion ($|x| \to \infty$) is valid:

$$\psi_\beta(x, \lambda) = \psi_\beta^0(x, \lambda) + R^0_{\lambda+i0}(x, 0) \sum_{n=0}^{N} \frac{1}{n!}\left(\frac{1}{2i\sqrt{\lambda}|x|}\right)^n B_n(\Delta_\alpha) f_\lambda(\beta, \alpha) + \psi_\beta^N(x, \lambda). \tag{1.11}$$

In this asymptotic formula $f_\lambda(\beta, \alpha)$ is the so-called scattering amplitude

$$f_\lambda(\beta, \alpha) \equiv -\int dy\, \psi_\beta(y, \lambda) q(y) \psi^0_{-\alpha}(y, \lambda). \tag{1.12}$$

The quantity $\psi_\beta^N(x,\lambda)$ is of the form

$$\psi_\beta^N(x, \lambda) = R^0_{\lambda+i0}(x, 0)\left(\frac{1+\sqrt{\lambda}}{|x|}\right)^{N+1} O_N(1), \tag{1.13}$$

where $O_N(1)$ is uniformly bounded with respect to x, λ, β and has bounded derivatives with respect to x and λ. If $q(x) = 0$ for $|x| \geq R$, then with $|x| > R$, the asymptotic expansion (1.11) converges to the function $\psi_\beta(x, \lambda)$ and $\psi_\beta^N(x,\lambda) \xrightarrow[N\to\infty]{} 0$.

Proof. The proof of this theorem follows from Lemma 1, the estimate

$$|\psi_\beta(x, \lambda)| \leqslant C(q) \tag{1.14}$$

and the equation

$$\psi_\beta(x, \lambda) = \psi_\beta^0(x, \lambda) - \int dy R^0_{\lambda+i0}(x, y) q(y) \psi_\beta(y, \lambda). \tag{1.15}$$

Let us now give a few definitions. Let us consider a unit sphere S in E_3. The points on the sphere will be denoted by the unit vectors $\alpha, \beta, \ldots$ Let K(S) be the space of infinitely differentiable basis functions $F(\alpha)$ in S defined in the usual manner. For these functions we define the Laplacian operator Δ_α in the same way as was done in Theorem 1. Let us consider the generalized functions $\varphi(F) \equiv (\varphi, F)$ in the space K(S) and the Laplace operator for them:

$$(\Delta_\alpha \varphi, F) \equiv (\varphi, \Delta_\alpha F).$$

We will require that the function $\psi_\beta^0(x,\lambda)$ be interpreted as a function in K'(S) of the variable $\alpha = x/|x|$,

$$(\psi_\beta^0(x, \lambda), F) \equiv \int e^{i\sqrt{\lambda}|x|\alpha\beta} F(\alpha)\, d\alpha.$$

The integration is over S, $d\alpha$ is an invariant element of surface of the unit sphere (in a spherical coordinate system $d\alpha = \sin\theta\, d\theta\, d\varphi$). Let us also introduce one more conventional symbol $\delta(\alpha - \beta)$:

$$\int \delta(\alpha - \beta) F(\alpha)\, d\alpha = F(\beta).$$

Lemma 2

The following equality is valid in K'(S) for $\sqrt{\lambda}|x| \to \infty$:

$$\psi_\beta^0(x,\lambda)=\frac{2\pi\cdot 4\pi}{i\sqrt{\lambda}}\frac{e^{i\sqrt{\lambda}|x|z}}{4\pi|x|}\sum_{n=0}^{N}\frac{1}{n!}\left(\frac{1}{2i\sqrt{\lambda}|x|}\right)^n z^n B_n(\Delta_\alpha)\delta(\alpha-\beta z)\Bigg|_{z=-1}^{z=+1}+\psi_\beta^{0,N}(x,\lambda). \tag{1.16}$$

Here, we have $\alpha = x/|x|$ and

$$|(\psi_\beta^{0,N}(x,\lambda),F)|<\frac{C_N}{(\sqrt{\lambda}|x|)^{N+2}}\|F\|_{C_{3(N+1)}(S)}\,. \tag{1.17}$$

$\|F\|_{C_M}$ is taken to mean the sum of the maximum moduli of all derivatives of the function F to order M.* The derivative $[\partial/(\partial\sqrt{\lambda}\,|x|)]\psi_\beta^{0,N}(x,\lambda)$ is also estimated by formula (1.17).

Proof. Let us introduce on S a spherical coordinate system θ,φ with its pole at the point β and let us take $z=\cos\theta$, then

$$(\psi_\beta^0(x,\lambda),F)=\int_{-1}^{1}dz\,e^{i\sqrt{\lambda}|x|z}\int_0^{2\pi}F(\alpha)\,d\varphi.$$

Integrating the outer integral by parts, we obtain an expansion of type (1.16) to within a constant factor. In doing this, we can easily obtain estimate (1.17) and the second proposition of the lemma. We also establish the possibility of the differentiation of asymptotic formula (1.16) with respect to the coordinates of the point x. After this, as in Lemma 1, the explicit expression for the coefficients is obtained from the equation

$$(-\Delta_x-\lambda)\,\psi_\beta^0(x,\lambda)=0.$$

Theorem 2.2

The following asymptotic expansion is valid for $\lambda>0$ as $|x|\to\infty$:

$$\psi_\beta(x,\lambda)\sim\frac{2\pi\cdot 4\pi}{i\sqrt{\lambda}}R^0_{\lambda+i0}(x,0)\sum_{n=0}^{\infty}\frac{1}{n!}\left(\frac{1}{2i\sqrt{\lambda}|x|}\right)^n B_n(\Delta_\alpha)S_\lambda(\beta,\alpha)+\overline{\frac{2\pi\cdot 4\pi}{i\sqrt{\lambda}}R^0_{\lambda+i0}(x,0)\sum_{n=0}^{\infty}\frac{1}{n!}\left(\frac{1}{2i\sqrt{\lambda}|x|}\right)^n B_n(\Delta_\alpha)\delta(\alpha+\beta)}. \tag{1.18}$$

Here, again, $\alpha = x/|x|$; the bar denotes a complex conjugate. $S_\lambda(\beta,\alpha)$ denotes the kernel of the "scattering matrix" — a unitary operator in $L_2(S)$

$$(S(\lambda)F)(\beta)=\int d\alpha\,S_\lambda(\beta,\alpha)\,F(\alpha), \tag{1.19}$$

which is defined by the formula

$$S_\lambda(\beta,\alpha)=\delta(\beta-\alpha)+\frac{i\sqrt{\lambda}}{2\pi\cdot 4\pi}f_\lambda(\beta,\alpha). \tag{1.20}$$

* The sphere is assumed to be covered by a finite number of simple regions in each of which a smooth nondegenerate coordinate grid is introduced. The derivatives are calculated with respect to the coordinate lines.

The convergence to the asymptotic formula is to be understood as convergence in K'(S).

The proof can be easily obtained from Theorem 2.1 in Lemma 2.

With this, we conclude our discussion of asymptotic behavior at large distances and we now proceed to the study of the behavior of $R_\lambda(x, x')$, $\psi_\alpha(x, \lambda)$, and $f_\lambda(\beta, \alpha)$ as $|\lambda| \to \infty$.

Theorem 3

As $|\lambda| \to \infty$ $(+0 \le \arg\lambda \le 2\pi - 0)$, the function $R_\lambda(x, x')$ possesses the following asymptotic expansion:

$$R_\lambda(x, x') = R_\lambda^0(x, x') \sum_{n=0}^{N} (-1)^n \left(\frac{1}{2i\sqrt{\lambda}}\right)^n \sum_{m=0}^{n} |x - x'|^m \times \\ \times \Omega_n^{(m)}(x, x') + \widetilde{R}_\lambda^N(x, x'). \tag{1.21}$$

The coefficients $\Omega_n^{(m)}(x, x')$ can be found from the recurrence relations

$$\Omega_{n+1}^{(m+1)}(x, x') = 2(m+1)\,\Omega_n^{(m+2)}(x, x') + \\ + \frac{1}{2}\int_{-1}^{1} d\eta \left(\frac{1}{2}(1-\eta)\right)^m \{\Delta_y \Omega_n^{(m)}(y, x') - q(y)\,\Omega_n^{(m)}(y, x') - \\ - m(m+1)\,\Omega_n^{(m+2)}(y, x')\}, \tag{1.22}$$

where $y = \frac{1}{2}(1-\eta)x + \frac{1}{2}(1+\eta)x'$ $(n = 0, 1, \ldots;\ m = 0, 1, \ldots, n)$; $\Omega_0^{(0)} = 1$, and, moreover, $\Omega_{n+1}^{(0)} = 0$ $(n = 0, 1, \ldots)$.

It follows from these formulas that $\Omega_{2s}^{2t-1} = \Omega_{2s-1}^{2t} = 0$ $(s = 1, 2, \ldots;\ t = 0, 1, \ldots)$. The remainder $\widetilde{R}_\lambda^N(x, x')$ is estimated as follows:

$$|\widetilde{R}_\lambda^N(x, x')| \leqslant \frac{C_N}{(\sqrt{\lambda})^{N+1}}(1 + |x| + |x'|)^{M(N)}, \tag{1.23}$$

where $M(N) > 0$ is a finite number.

We will prove the following lemma before we proceed to the proof of Theorem 3.

Using the conventional notation, we introduce the space $S(E_3)$ of basis functions — infinitely differentiable functions decreasing as $|x| \to \infty$ together with their derivatives faster than any power of $|x|^{-1}$. We will consider $R_\lambda^0(x, y)\,R_\lambda^0(y, x')$ to be the generalized function in this space:

$$(R_\lambda^0(x, \cdot)\,R_\lambda^0(\cdot, x'),\ G) = \int dy\, \frac{e^{i\sqrt{\lambda}|x-y|}}{4\pi|x-y|} \cdot \frac{e^{i\sqrt{\lambda}|y-x'|}}{4\pi|y-x'|}\, G(y). \tag{1.24}$$

Lemma 3

The following relation holds as $|\lambda| \to \infty$ $(+0 \le \arg\lambda \le 2\pi - 0)$:

$$R_\lambda^0(x, y)\,R_\lambda^0(y, x') = \\ = R_\lambda^0(x, x') \sum_{n=0}^{N} (-1)^{n+1} \left(\frac{1}{2i\sqrt{\lambda}}\right)^{n+1} \sum_{m=1}^{n+1} |x - x'|^m \Phi_n^{(m)}(x, x' | y) + \Gamma_\lambda^N(x, x' | y). \tag{1.25}$$

The coefficients $\Phi_n^{(m)}$ are defined by the recurrence relations

$$\Phi_0^{(1)}(x, x' \mid y) = \frac{1}{2}\int_{-1}^{1} d\eta\,\delta(z-y), \text{ and } \Phi_{n+1}^{(m)}(x, x' \mid y) = 2m\Phi_n^{(m+1)}(x, x' \mid y) + \frac{1}{2}\int_{-1}^{1} d\eta \times$$
$$\times \left(\frac{1}{2}(1-\eta)\right)^{m-1}\left[\Delta_z \Phi_n^{(m-1)}(z, x' \mid y) - m(m-1)\Phi_n^{(m-1)}(z, x' \mid y)\right], \tag{1.26}$$

where $z = \frac{1}{2}(1-\eta)x + \frac{1}{2}(1+\eta)x'$ and $\delta(y)$ denotes the delta function. The following estimate holds:

$$\left|(\Gamma_\lambda^N(x, x' \mid \cdot), G)\right| \leqslant C_N(G)\frac{\left|e^{i\sqrt{\lambda}|x-x'|}\right|}{(\sqrt{\lambda})^{N+2}}\frac{(1+|x-x'|)^{2(N+1)}}{(1+|x|)(1+|x'|)}. \tag{1.27}$$

Proof of the Lemma. Let us consider the integral (1.24). With x and x' fixed, we introduce a Cartesian coordinate system $y = \{y_1, y_2, y_3\}$ with its center at the point $(x + x')/2$, and axis y_3 directed toward the point x. In the integral (1.24) we make a change of variable to the elongated ellipsoidal coordinates

$$\xi = \frac{1}{2}(|y-x'| + |x-y|), \quad \eta = \frac{|y-x'| - |y-x|}{|x-x'|},$$

where φ is the angle between the y_1 axis and the projection of y onto the plane $\{y_1, y_2\}$. In terms of the new variables, the integral (1.24) becomes

$$\left(\frac{1}{4\pi}\right)^2 \int_{\frac{1}{2}|x-x'|=d}^{\infty} d\xi e^{2i\sqrt{\lambda}\xi}\int_{-1}^{1} d\eta \int_0^{2\pi} d\varphi \times$$
$$\times G\left(\cos\varphi\sqrt{(1-\eta^2)(\xi^2-d^2)}\,\sin\varphi\sqrt{(1-\eta^2)(\xi^2-d^2)},\ \xi\eta\right).$$

Integrating by parts, we obtain to within the description of the coefficients of $\Phi_n^{(m)}$ the expansion (1.25). Estimate (1.27) is obtained in the course of this. After this, we make use of the equation

$$(-\Delta_x - \lambda)R_\lambda^0(x, y)R_\lambda^0(y, x') = \delta(x-y)R_\lambda^0(y, x'),$$

from which it follows that

$$\delta(x-y) = \Phi_0^{(1)} + (x-x', \nabla_x\Phi_0^{(1)})$$

and

$$\left(x-x', \nabla_x\left[\Phi_{n+1}^{(m)} - 2m\Phi_n^{(m+1)}\right]\right) + m\left[\Phi_{n+1}^{(m)} - 2m\Phi_n^{(m+1)}\right] =$$
$$= \Delta_x\Phi_n^{(m-1)} - m(m-1)\Phi_n^{(m+1)}.$$

Since the coefficients $\Phi_n^{(m)}(x, x' \mid y)$ must generate functionals depending smoothly on the parameters x and x', these equations yield the recurrence formulas of the lemma.

Proof of Theorem 3. All of the assertions of Theorem 3 except the recurrence formulas follow from Lemma 3, integral equation (1.9), if the latter is integrated a sufficient number of times, and estimate (1.10). The recurrence formulas for the coefficients $\Omega_n^{(m)}$ are obtained with the help of the differential equation

$$[-\Delta_x + q(x) - \lambda] R_\lambda(x, x') = \delta(x - x').$$

Theorem 4

As $\lambda \to +\infty$,

$$\psi_\alpha(x, \lambda) = \psi_\alpha^0(x, \lambda) \sum_{n=0}^{N} (-1)^n \left(\frac{1}{2i\sqrt{\lambda}}\right)^n \omega_n(x, \alpha) + \tilde{\psi}_\alpha^N(x, \lambda), \tag{1.28}$$

where $\omega_0(x, \alpha) = 1$ and

$$\omega_{n+1}(x, \alpha) = \int_{-\infty}^{0} dt\, [\Delta_y \omega_n(y, \alpha) - q(y)\omega_n(y, \alpha)], \tag{1.29}$$

where $y = x + \alpha t$.

The quantity $\tilde{\psi}_\alpha^N(x, \lambda)$ can be estimated as

$$|\tilde{\psi}_\alpha^N(x, \lambda)| \leq \frac{C_N}{(\sqrt{\lambda})^{N+1}} (1 + |x|)^{M(N)}. \tag{1.30}$$

Proof. The proof follows from the integral equation (1.15) which should be integrated a sufficient number of times, estimate (1.16), Lemma 3 and its analog in which one of the factors of type $R_\lambda^0(x, y)$ is replaced by the plane wave $\psi_\alpha^0(y, \lambda)$, and the differential equation

$$[-\Delta_x + q(x) - \lambda] \psi_\alpha(x, \lambda) = 0.$$

Theorem 5

The scattering amplitude $f_\lambda(\beta, \alpha)$ has the following asymptotic expansion as $\lambda \to +\infty$:

$$f_\lambda(\beta, \alpha) \sim -\sum_{n=0}^{\infty} (-1)^n \left(\frac{1}{2i\sqrt{\lambda}}\right)^n \int dy\, q(y)\, \omega_n(y, \beta)\, \psi_\lambda^0(y, \beta - \alpha). \tag{1.31}$$

The proof is obvious on the basis of definition (1.12) and Theorem 4.

Note. The integral $\int dy\, q(y)\, \omega_n(y, \beta)\psi_\lambda^0(y, \beta - \alpha)$ with $\alpha = -\beta$ decreases faster than $1/\lambda$ as $\lambda \to +\infty$.

§ 2. Calculation of $\mathrm{Sp}(R_\lambda - R_\lambda^0)$

It is known that with our assumptions concerning the potential $q(x)$, the operators H and H_0 are resolvent comparable; in other words, outside the spectral points of the operator H there exists a trace $\mathrm{Sp}(R_\lambda - R_\lambda^0)$ analytically dependent on λ.

Theorem 6

With $\mathrm{Im}\,\lambda \neq 0$, the relation

$$\mathrm{Sp}(R_\lambda - R_\lambda^0) = \int_{-\infty}^{0} \frac{1}{\mu - \lambda}\, d \sum_{\substack{l \\ \mu > \lambda_l}} m_l + \frac{1}{2\pi i} \int_0^{\infty} d\mu \frac{1}{\mu - \lambda} \frac{d}{d\mu} \ln \det S(\mu) \tag{2.1}$$

holds and

$$\lim_{\varepsilon\downarrow 0} \operatorname{Im} \operatorname{Sp}(R_{\lambda+i\varepsilon} - R^0_{\lambda+i\varepsilon}) = \frac{1}{2i}\cdot\frac{d}{d\lambda}\ln\det S(\lambda) \quad (\lambda > 0), \tag{2.2}$$

where $S(\mu)$ is the scattering matrix.

Lemma 4

Let B and C be integral operators of the Hilbert–Schmidt type in L_2 with kernels $B(x, y)$ and $C(x, y)$. The integral operator $A = CB$

$$A(x, y) = \int dz C(x, z) B(z, y) \tag{2.3}$$

has an absolute trace and

$$\operatorname{Sp} A = \int dx A(x, x). \tag{2.4}$$

Proof of the Lemma. Only formula (2.4) needs a proof. Let $\{\varphi_n(x)\}$ be a real orthonormal basis in L_2; then,

$$\operatorname{Sp} A = \sum_n (A\varphi_n, \varphi_n)_{L_2} = \sum_n (B\varphi_n, C^*\varphi_n)_{L_2} =$$
$$= \sum_{n,m} (B\varphi_n, \varphi_m)_{L_2} (\varphi_m, C^*\varphi_n)_{L_2} = \sum_{n,m} (B, \varphi_m \times \varphi_n)_{L_2\times L_2} \times$$
$$\times (C, \varphi_n \times \varphi_m)_{L_2\times L_2} = \iint dx dz C(x, z) B(z, x) = \int dx A(x, x).$$

In these formulas, $\varphi_n \times \varphi_m$ is an element of $L_2 \times L_2$: $\varphi_n(x)\varphi_m(y)$. In transposing the order of summations, we have made use of the absolute convergence of the series, while in the last equality we have used Fubini's theorem, whose validity in this case is ensured by the absolute convergence of the repeated integral $\int dxA(x, x)$:

$$\int_{|x|<N} dx \int_{|z|<N'} dz\, |C(x, z) B(z, x)| \leqslant$$
$$\leqslant \iint_{\substack{|x|<N \\ |z|<N'}} dx dz\, |C(x, z) B(z, x)| \leqslant \|C\|_{L_2\times L_2} \|B\|_{L_2\times L_2}.$$

The lemma has been proved.

Proof of Theorem 5. Lemma 4 allows us to calculate $\operatorname{Sp}(R_\lambda - R^0_\lambda)$ from the formula

$$\operatorname{Sp}(R_\lambda - R^0_\lambda) = \int dx\, [R_\lambda(x, x') - R^0_\lambda(x, x')]\,|_{x=x'}, \tag{2.5}$$

since, in view of Eq. (1.9) the kernel $R_\lambda(x, x') - R^0_\lambda(x, x')$ is the convolution of two Hilbert–Schmidt kernels

$$R_\lambda(x, x')\,|q(x')|^{\frac{1}{2}} \quad \text{and} \quad \operatorname{sign} q(x)\,|q(x)|^{\frac{1}{2}} R^0_\lambda(x, x').$$

Let us make use of the spectral representation of the resolvent R_λ

$$R_\lambda(x, x') = \int_{-\infty}^{0} \frac{1}{\mu-\lambda}\, d \sum_{\substack{l \\ \mu>\lambda_l}} P_l(x, x') +$$
$$+ \int_0^\infty \frac{1}{\mu-\lambda}\, d\left\{\int_0^\mu d\mu\, \frac{\sqrt{\mu}}{2}\left(\frac{1}{2\pi}\right)^3 \int d\beta\, \psi_\beta(x, \mu)\, \overline{\psi_\beta(x', \mu)} + \sum_{l=1}^{M} P_l(x, x')\right\}.$$

Here, $P_l(x,x')$ is the kernel of an integral operator, the projection on the m_l-dimensional characteristic subspace H corresponding to the eigenvalue λ_l; $\psi_\beta(x, \mu)$ is the solution of the scattering problem. Subtracting from this the analogous representation of $R^0_\lambda(x,x')$ and applying formula (2.5), we obtain

$$\mathrm{Sp}(R_\lambda - R^0_\lambda) = \int_{-\infty}^{0} \frac{1}{\mu-\lambda}\, d \sum_{\substack{l \\ \mu>\lambda_l}} m_l + \lim_{R\to\infty} \int_0^\infty d\mu\, \frac{1}{\mu-\lambda}\cdot\frac{\sqrt{\mu}}{2}\left(\frac{1}{2\pi}\right)^3 \int d\beta \int_{|x|\leqslant R} dx\, [\,|\psi_\beta(x, \mu)|^2 - 1]. \tag{2.6}$$

Let us now consider the integral

$$\int_{|x|\leqslant R} dx\, [\,|\psi_\beta(x, \mu)|^2 - 1]. \tag{2.7}$$

In the following, we omit a number of details concerning the behavior of integral (2.7) as $R\to\infty$ because of lack of space. Since the function $\psi_\beta(x,\mu)$ satisfies the equation

$$[-\Delta_x + q(x) - \mu]\, \psi_\beta(x, \mu) = 0,$$

the integral (2.7) can be transformed as follows:

$$\int_{|x|\leqslant R} dx\, [\,|\psi_\beta(x, \mu)|^2 - 1] =$$
$$= R^2 \frac{\partial}{\partial \nu}\bigg|_{\nu=\mu} \int d\alpha \left\{\psi_\beta(R\alpha, \nu)\, \frac{\overleftrightarrow{\partial}}{\partial R}\, \overline{\psi_\beta(R\alpha, \mu)} - \right.$$
$$\left. - \psi^0_\beta(R\alpha, \nu)\, \frac{\overleftrightarrow{\partial}}{\partial R}\, \overline{\psi^0_\beta(R\alpha, \mu)}\right\}. \tag{2.8}$$

The symbol $\partial/\partial\nu\big|_{\nu=\mu}$ denotes differentiation with respect to ν after which ν is set equal to μ;

$$\psi^{(1)} \frac{\overleftrightarrow{\partial}}{\partial R} \psi^{(2)} \equiv \psi^{(1)} \frac{\partial}{\partial R} \psi^{(2)} - \psi^{(2)} \frac{\partial}{\partial R} \psi^{(1)}.$$

Let us substitute the asymptotic representation of the function $\psi_\beta(R\sigma, \mu)$ as $R\to\infty$ (Theorem 2.1) into integral (2.8). Neglecting terms that tend to zero as $R\to\infty$, we can retain in the sums appearing in (1.11) the three leading terms. Referring now to Lemma 2, we limit ourselves to the two leading terms. After this, all of the plane waves in the first and second terms of formula (2.8) can be replaced simultaneously by the asymptotic expression of Lemma 2 in which only the two leading terms are retained. This means that as $R\to\infty$ the functions ψ and ψ^0 on the right-hand side of (2.8) can be replaced by the two leading terms of the asymptotic expansion of Theorem 2.2. Further calculations show that because of cancellation, the second-order terms in this expansion do not give a contribution to integral (2.8) as $R\to\infty$. Therefore, we must restrict ourselves to the leading terms of the asymptotic expansion of Theorem 2.2. These terms no longer give contributions that tend to zero as $R \to \infty$, but many of the terms generated by them cancel out because of the unitary nature of the scattering matrix $S(\mu)$. The final result is

$$\int\limits_{|x|<R} dx\,[|\psi_\beta(x,\mu)|^2-1] \sim \\ \sim \frac{1}{2\pi i}\cdot\frac{2}{\sqrt{\mu}}(2\pi)^3\int d\alpha\,\overline{S_\mu(\beta,\alpha)}\cdot\frac{\partial}{\partial\mu}S_\mu(\beta,\alpha) - \\ -\frac{1}{4\mu}\left[e^{2i\sqrt{\mu}R}f_\mu(\beta,-\beta)+e^{-2i\sqrt{\mu}R}\overline{f_\mu(\beta,-\beta)}\right]. \tag{2.9}$$

After the substitution of this result into formula (2.6) for $\mathrm{Sp}(R-R^0_\lambda)$, the second term yields a zero contribution according to the Riemann–Lebesgue lemma. Its validity is ensured by Theorem 5 and the accompanying remark.

We thus have

$$\mathrm{Sp}(R_\lambda - R^0_\lambda) = \int\limits_{-\infty}^{0}\frac{1}{\mu-\lambda}\,d\sum_{\substack{l\\ \mu>\lambda_l}} m_l + \\ +\frac{1}{2\pi i}\int\limits_0^\infty d\mu\,\frac{1}{\mu-\lambda}\int d\beta\int d\alpha\,\overline{S_\mu(\beta,\alpha)}\frac{\partial}{\partial\mu}S_\mu(\beta,\alpha). \tag{2.10}$$

It is easy to see that

$$\int d\beta\int d\alpha\,\overline{S_\mu(\beta,\alpha)}\frac{\partial}{\partial\mu}S_\mu(\beta,\alpha) = \frac{d}{d\mu}\ln\det S(\mu). \tag{2.11}$$

This is established as follows. $(\partial/\partial_\mu)S_\mu(\beta,\alpha)$ is the kernel of the operator $(d/d_\mu)S(\mu)$. This kernel is a smooth (infinitely differentiable) function. The Hermitian and anti-Hermitian parts of the kernel possess the same property. The last two operators possess traces as follows from the asymptotic estimates for the eigenvalues of integral operators (for example, see [21]). Consequently, an abstract trace exists in the case of operators $(d/d\mu)S(\mu)$ and $S^*(\mu)(d/d\mu)S(\mu)$. Since the kernel of the last operator is again a smooth function, it can be calculated as the trace of the kernel [22], so that

$$\int d\beta\int d\alpha\,\overline{S_\mu(\beta,\alpha)}\frac{\partial}{\partial\mu}S_\mu(\beta,\alpha) = \mathrm{Sp}\,S^*(\mu)\frac{d}{d\mu}S(\mu).$$

The right-hand side transforms as follows:

$$\mathrm{Sp}S^*(\mu)\frac{d}{d\mu}S(\mu) = \frac{d}{d\mu}\mathrm{Sp}\ln S(\mu) = \frac{d}{d\mu}\ln\det S(\mu).$$

Here, we have used the existence of $\mathrm{Sp}[S(\mu)-I]$, which is established in the same way as we have earlier established the existence of $\mathrm{SP}(d/d\mu)S(\mu)$ (see [15] for the corresponding abstract assertion).

The substitution of (2.11) into formula (2.10) gives the first assertion of Theorem 6. The second proposition is a consequence of the first based on the following property of the trace:

$$\operatorname{Sp}(R_{\bar{\lambda}} - R^0_{\bar{\lambda}}) = \overline{\operatorname{Sp}(R_\lambda - R^0_\lambda)}. \tag{2.12}$$

The theorem has been proved.

Let us list some of the properties of the function $\operatorname{Sp}(R_\lambda - R^0_\lambda)$ needed in the following:

1) $\operatorname{Sp}(R_\lambda - R^0_\lambda)$ is an analytic function in the λ plane with the exception of the half-axis $\lambda \geq 0$ and the points λ_l at which it has simple poles with residues m_l [this follows from (2.1)];

2) when λ tends to the points of the half-axis $\lambda > 0$ with the sign of $\operatorname{Im}\lambda$ remaining unchanged, then $\operatorname{Sp}(R_\lambda - R^0_\lambda)$ uniformly assumes continuous and bounded ($\lambda > 0$) limiting values [this follows from (2.1)];

3) as $\lambda \to 0$, $\operatorname{Sp}(R_\lambda - R^0_\lambda)$ may tend to infinity not faster than $\lambda^{-1/2}$.

The last assertion can be proved with the help of formula (2.1) if it is noted that $(d/d\lambda) \cdot \ln\det S(\lambda)$ behaves in an analogous manner for real λ. The fact is that the scattering amplitude $f_\lambda(\beta,\alpha)$ in terms of which we express $(d/d\lambda)\ln\det S(\lambda)$ with the help of formula (2.11) is a smooth function of $\sqrt{\lambda}$ and it can be differentiated with respect to $\sqrt{\lambda}$ any number of times when $\lambda \geq 0$. This is a consequence of the same behavior of the function $\psi_\beta(x,\lambda)$ which is ensured by the assumptions made in § 1 concerning the solutions of Eq. (1.2).

Theorem 7

The following asymptotic expression uniform over $\arg\lambda$ holds as $|\lambda| \to \infty$:

$$\operatorname{Sp}(R_\lambda - R^0_\lambda) \sim \frac{1}{2i\sqrt{\lambda}} \cdot \frac{1}{4\pi} \int dx\, q(x) - \frac{1}{2i\sqrt{\lambda}\,\lambda} \sum_{l=0}^{\infty} \frac{2l+1}{\lambda^l} Q_l, \tag{2.13}$$

where

$$Q_l = \frac{1}{2l+1} \cdot \frac{(-1)^{l+1}}{4^{l+1}} \cdot \frac{1}{4\pi} \int dx\, \Omega^{(1)}_{2l+3}(x, x). \tag{2.14}$$

<u>Proof.</u> We have to show that the asymptotic expression of Theorem 3 for $R_\lambda(x,x') - R^0_\lambda(x,x')\big|_{x=x'}$ can be integrated term by term with respect to x. Let us write Eq. (1.9) in the following manner:

$$R_\lambda - R^0_\lambda = -R^0_\lambda V R_\lambda. \tag{2.15}$$

It follows from this that

$$R_\lambda - R^0_\lambda = -R^0_\lambda V R^0_\lambda + R^0_\lambda V R_\lambda V R^0_\lambda. \tag{2.16}$$

Let us take the trace of this equality. After some simple calculations, we obtain

$$\operatorname{Sp}(R_\lambda - R^0_\lambda) = \frac{1}{2i\sqrt{\lambda}} \cdot \frac{1}{4\pi} \int dy\, q(y) - \frac{1}{2i\sqrt{\lambda}} \cdot \frac{1}{4\pi} \int dy\,dz\, e^{i\sqrt{\lambda}|y-z|} q(y)\, q(z)\, R_\lambda(y, z). \tag{2.17}$$

This formula allows us to obtain easily an estimate of the error after the term-by-term integration of the asymptotic expression over x. Let us write the asymptotic expression of Theorem 3 as follows:

$$R_\lambda(x, x') = \widetilde{Q}^N_\lambda(x, x') + \widetilde{R}^N_\lambda(x, x'). \tag{2.18}$$

The contribution to $\mathrm{Sp}(R - R_\lambda^0)$ from $\widetilde{R}_\lambda^N(x,x')$ is of the order $(\sqrt{\lambda})^{-N-2}$. Let us calculate the contribution from $\widetilde{Q}_\lambda^N(x, x')$. It should be noted that

$$q(x)\widetilde{Q}_\lambda^N(x, x') = -(-\Delta_x - \lambda)\widetilde{Q}_\lambda^N(x, x') + O\left((\sqrt{\lambda})^{-N}\right).$$

Using this relation and inverting the calculations which led from (2.15) to (2.16), we obtain

$$\mathrm{Sp}(R_\lambda - R_\lambda^0) = \int dx\left[\widetilde{Q}_\lambda^N(x, x') - R_\lambda^0(x, x')\right]_{x=x'} + O\left((\sqrt{\lambda})^{-N-1}\right), \tag{2.19}$$

which was to be proved.

§ 3. Trace Formulas

Lemma 5

Let the function $f(\lambda)$ of a complex variable λ possess the following properties: 1) $f(\lambda)$ is regular in the plane λ with the exception of the half-axis $\lambda \geq 0$ and the points λ_l ($\widetilde{\lambda}_l < 0$, $l = 1, 2, \ldots, \widetilde{M} < \infty$), where it has simple poles with residues $\widetilde{m}_l$; 2) if λ tends to the points of the half-axis $\lambda \geq 0$, the sign of $\mathrm{Im}\,\lambda$ being maintained, then $f(\lambda)$ uniformly assumes continuous and bounded ($\lambda > 0$) limiting values; 3) $f(\lambda + i0) = f(\lambda - i0)$ ($\lambda > 0$); 4) $f(\lambda)$ can become infinite when $\lambda \sim 0$ no faster than $\lambda^{-1/2}$; 5) as $|\lambda| \to \infty$, the following asymptotic expression uniform over $\arg\lambda$ is valid:

$$f(\lambda) \sim \frac{-1}{2i\sqrt{\lambda}\,\lambda}\sum_{l=0}^{\infty}\frac{2l+1}{\lambda^l}\widetilde{Q}_l, \tag{3.1}$$

where $\widetilde{Q}_l$ is real.

Let us write

$$\eta(\lambda) \equiv \int_\lambda^\infty d\lambda\,\mathrm{Im} f(\lambda + i0) \quad (\lambda > 0). \tag{3.2}$$

Then, we have

a)

$$\sum_{l=1}^{\widetilde{M}} \widetilde{m}_l = -\frac{1}{\pi}\eta(0), \tag{3.3}$$

b) with $\mu = 1, 2, \ldots$

$$\sum_{l=1}^{\widetilde{M}} \widetilde{m}_l \widetilde{\lambda}_l^\mu + \frac{\mu}{\pi}\int_0^\infty d\lambda\,\lambda^{\mu-1}\left[\eta(\lambda) - \frac{1}{\sqrt{\lambda}}\sum_{l=0}^{\mu-1}\frac{\widetilde{Q}_l}{\lambda^l}\right] = 0. \tag{3.4}$$

Proof. Let us consider the integral

$$\frac{1}{2\pi i}\oint_{C_N}\lambda^s f(\lambda)\,d\lambda = -\sum_{l=1}^{\widetilde{M}}\widetilde{m}_l\widetilde{\lambda}_l^s. \tag{3.5}$$

Here $\lambda^s = e^{s\ln\lambda}$ and $\ln\lambda$ is defined on the plane with a cut $\lambda \geq 0$ by the selection of the principal branch. The contour C_N consists of two sections, one of them contains the segment $[0, N]$, the other a circle of radius N. Formula (3.5) is valid for sufficiently large N. If $-\frac{1}{2} < \mathrm{Re}\, s < \frac{1}{2}$, then, assuming that N → ∞ in this formula and deforming the contour section containing the cut to the half-axis [0, ∞), we obtain

$$e^{i\pi s}\cos\pi s\,\frac{1}{\pi}\int_0^\infty d\lambda\lambda^s\,\mathrm{Im}\,f(\lambda+i0)-e^{i\pi s}\sin\pi s\,\frac{1}{\pi}\int_0^\infty d\lambda\lambda^s\,\mathrm{Re}\,f(\lambda)=$$

$$=-\sum_{l=1}^{\widetilde{M}}\widetilde{m}_l\widetilde{\lambda}_l^s. \tag{3.6}$$

Formula (3.3) of the lemma follows from this with s = 0.

Let us now consider the analytic continuation of equality (3.6) with respect to the variable s from the strip $-\frac{1}{2} < \mathrm{Re}\, s < \frac{1}{2}$ to the region $\mathrm{Re}\, s \geq \frac{1}{2}$. The integral

$$\int_0^\infty d\lambda\lambda^s\,\mathrm{Re}\,f(\lambda)$$

is analytically continued as a regular function by its own formula, so that for $\mu = 1, 2, \ldots$ we have

$$\left[\text{Analytical continuation of }\frac{1}{\pi}\int_0^\infty d\lambda\lambda^s\,\mathrm{Im}\,f(\lambda+i0)\right]_{s=\mu}=-\sum_{l=1}^{\widetilde{M}}\widetilde{m}_l\widetilde{\lambda}_l^\mu. \tag{3.7}$$

Let us evaluate the left-hand side of this equality. For $-\frac{1}{2} < \mathrm{Re}\, s < \frac{1}{2}$ we have

$$\frac{1}{\pi}\int_0^\infty d\lambda\lambda^s\,\mathrm{Im}\,f(\lambda+i0)=\frac{1}{\pi}\int_0^1 d\lambda\lambda^s\,\mathrm{Im}\,f(\lambda+i0)+$$

$$+\frac{1}{\pi}\int_1^\infty d\lambda\lambda^s\left[\mathrm{Im}\,f(\lambda+i0)-\frac{1}{2\sqrt{\lambda}\lambda}\sum_{l=0}^{\mu-1}\frac{2l+1}{\lambda^l}\widetilde{Q}_l\right]-$$

$$-\frac{1}{2\pi}\sum_{l=0}^{\mu-1}\frac{2l+1}{s-l-\frac{1}{2}}\widetilde{Q}_l.$$

Both integrals obtained can be analytically continued by their formulas to the point $s = \mu$. Let us now note that

$$-\frac{1}{2\pi}\sum_{l=0}^{\mu-1}\frac{2l+1}{\mu-l-\frac{1}{2}}\widetilde{Q}_l=-\frac{1}{\pi}\int_0^1 d\lambda\lambda^\mu\,\frac{1}{2\sqrt{\lambda}}\cdot\frac{1}{\lambda}\sum_{l=0}^{\mu-1}\frac{2l+1}{\lambda^l}\widetilde{Q}_l,$$

hence

$$\left[\text{Analytical continuation of }\frac{1}{\pi}\int_0^\infty d\lambda\lambda^s\,\mathrm{Im}\,f(\lambda+i0)\right]_{s=\mu}=\frac{1}{\pi}\int_0^\infty d\lambda\lambda^\mu\left[\mathrm{Im}\,f(\lambda+i0)-\frac{1}{2\sqrt{\lambda}\lambda}\sum_{l=0}^{\mu-1}\frac{2l+1}{\lambda^l}\widetilde{Q}_l\right]=$$

$$=\frac{\mu}{\pi}\int_0^\infty d\lambda\lambda^{\mu-1}\left[\eta(\lambda)-\frac{1}{\sqrt{\lambda}}\sum_{l=0}^{\mu-1}\frac{\widetilde{Q}_l}{\lambda^l}\right]. \tag{3.8}$$

If we now recall (3.7), we obtain formulas (3.4). The lemma has been proved.

The conditions of Lemma 5 are satisfied by the function

$$f(\lambda)=\mathrm{Sp}\,(R_\lambda-R_\lambda^0)-\frac{1}{2i\sqrt{\lambda}}\cdot\frac{1}{4\pi}\int dxq\,(x). \tag{3.9}$$

In this, we have $\tilde{\lambda}_l=\lambda_l$, $\tilde{m}_l=m_l$, $\tilde{M}=M$, $\tilde{Q}_l=Q_l$ (see Theorem 7), and

$$\begin{aligned}\gamma_1(\lambda)&=\int_\lambda^\infty d\mu\left[\frac{1}{2i}\cdot\frac{d}{d\mu}\ln\det S(\mu)+\frac{1}{2\sqrt{\mu}}\cdot\frac{1}{4\pi}\int dxq(x)\right]=\\&=-\frac{1}{2i}\left[\ln\det S(\lambda)+2i\sqrt{\lambda}\frac{1}{4\pi}\int dxq(x)\right].\end{aligned} \tag{3.10}$$

Lemma 5 thus leads to Theorem 8.

Theorem 8

ln det S(λ) satisfies the following system of identities:

$$\sum_{l=1}^{M}m_l=-\frac{1}{2\pi i}\ln\det S(0), \tag{3.11}$$

$$\sum_{l=1}^{M}m_l\lambda_l^\mu=\frac{\mu}{2\pi i}\int_0^\infty d\lambda\lambda^{\mu-1}\left[\ln\det S(\lambda)+2i\sqrt{\lambda}\frac{1}{4\pi}\int dxq\,(x)+2i\frac{1}{\sqrt{\lambda}}\sum_{l=0}^{\mu-1}\frac{Q_l}{\lambda^l}\right] \tag{3.12}$$

where $\mu=1,2,\ldots$

Identities (3.11) and (3.12) are called "trace formulas." The coefficients Q_l in these formulas can be expressed explicitly in terms of the potential q(x) with the help of Theorems 3 and 7. As an example, we give formula (3.12) for Theorem 8 with $\mu=1$. It can be easily shown from the recurrence relations of Theorem 3 that

$$\mathfrak{Q}_3^{(1)}(x,\ x)=q^2(x)-\frac{1}{2}\Delta q(x),\ Q_0=-\frac{1}{4}\cdot\frac{1}{4\pi}\int dxq^2(x).$$

After this, the trace formula (3.12) with $\mu=1$ takes the form

$$\sum_{l=1}^{M}m_l\lambda_l=\frac{1}{2\pi i}\int_0^\infty d\lambda\left[\ln\det S(\lambda)+2i\sqrt{\lambda}\frac{1}{4\pi}\int dxq\,(x)+\frac{1}{2i\sqrt{\lambda}}\cdot\frac{1}{4\pi}\int dxq^2(x)\right].$$

Literature Cited

1. V.S. Buslaev, Trace formulas for the three-dimensional Schroedinger operator, Dokl. Akad. Nauk SSSR, 143(5) (1962).
2. V.S Buslaev and L.D. Faddeev, Trace Formulas for the differential Sturm−Liouville operator, Dokl. Akad. Nauk SSSR, 132(1) (1960).

3. I.M. Gel'fand and B.M. Levitan, A simple identity for the eigenvalues of a second-order differential operator, Dokl. Akad. Nauk, 88(4) (1953).
4. L.A. Dikii, Trace formulas for the Sturm–Liouville differential operators, Usp. Matem. Nauk, 13(3) (1958).
5. M.I. Lomonosov, Differences of traces of one-dimensional operators defined over an infinite interval, Tr. ARTA, No. 31. Khar'kov (1957).
6. M.I. Lomonosov, Differences of traces of two Schroedinger operators, Tr. ARTA, No. 31. Khar'kov (1957).
7. N. Levinson, On the uniqueness of potential in a Schroedinger equation for a given asymptotic phase, Kgl. Danske. Videnskab Selskab, Mat.-Fys. Medd., 25(9) (1949).
8. R. Newton, Remarks on scattering theory, Phys. Rev., 101(6) (1956).
9. L.D. Faddeev, Expressions for the trace of the difference of two Sturm–Liouville singular differential operators, Dokl. Akad. Nauk SSSR, 115(5) (1957).
10. I. Percival, Energy moments of scattering phase shifts, Proc. Phys. Soc., 80(6) (1962).
11. I. Percival and M. Roberts, Energy moments of scattering phase shifts. II. Higher partial waves, Proc. Phys. Soc., 82(4) (1963).
12. M. Roberts, Energy moments and cross sections for a Gaussian potential, Proc. Phys. Soc., 82(4) (1963).
13. M.G. Krein, Trace formulas in perturbation theory, Matem. sb., 33(75), No. 3 (1953).
14. M.I. Lomonosov, The number of eigenvalues in the quantum scattering problem, Tr. ARTA, No. 63. Khar'kov (1962).
15. M.Sh. Birman and M.G. Krein, On the theory of wave operators and scattering operators, Dokl. Akad. Nauk SSSR, 144(3) (1962).
16. F.A. Berezin, The trace formula for the many-particle Schroedinger equation, Dokl. Akad. Nauk SSSR, 157(5) (1964).
17. A.Ya. Povzner, Expansion of an arbitrary function in terms of the eigenfunctions of the operator $\Delta u + qu$, Matem. sb., 32(74), No. 1 (1953).
18. A. Ya. Povzner, Expansions in terms of the eigenfunctions of the Schroedinger equation, Dokl. Akad. Nauk SSSR, 104(3) (1955).
19. T. Kato, Growth properties of solutions of the reduced wave equation with a variable coefficient, Comm. Pure and Appl. Math., 12(3) (1959).
20. T. Ikebe, Eigenfunction expansions associated with the Schroedinger operators and their applications to scattering theory. Arch. Rat. Mech. Anal., 5(1) (1960).
21. A.O. Gel'fond, Growth properties of the eigenvalues of homogeneous integral equations, Appendix to: W. V. Lovitt, Linear Integral Equations [Russian translation], GITTL, Moscow (1957).
22. I. Ts. Gokhberg and M. G. Krein, Introduction to the Theory of Linear Nonself-Adjoint Operators in Hilbert Space, Izd. Nauka (1966).

THE NONSELF-ADJOINT SCHROEDINGER OPERATOR

B. S. Pavlov

Introduction

The present paper is devoted to an investigation of the spectral properties of second-order nonself-adjoint differential operators. The first two sections contain a discussion of the operator l_h generated in $L_2(0, \infty)$ by the differential expression

$$ly = -y'' + q(x)y \tag{1}$$

and the boundary condition

$$y'(0) - hy(0) = 0. \tag{2}$$

Here, h is a complex number of q(x), a measurable complex function. Everywhere, unless specifically stated otherwise, we assume that the following condition holds:

$$\int_0^\infty x|q(x)|dx < \infty. \tag{3}$$

The spectral analysis of l_h has been performed by Naimark (see [1, 2, 3]). The main result of these papers is the proof of the theorem of expansion in eigenfunctions and adjoint functions of the operator l_h. Some results concerning the spectrum of the operator were also obtained. In particular, it was shown that the continuous spectrum of l_h fills the half-axis $[0, \infty)$, while the discrete spectrum consists of eigenvalues of finite rank. The set of eigenvalues is bounded and is at most denumerable. The eigenvalues can only accumulate to points of the half-axis $[0, \infty)$.

It should be noted that in the self-conjugate case $[\operatorname{Im} q(x) \equiv 0 = \operatorname{Im} h]$, condition (3) leads to the finiteness of the number of eigenvalues.* In the nonself-adjoint case, Naimark was able to prove the finiteness of the number of eigenvalues under the condition

$$\sup_x |q(x)| \exp(\varepsilon x) < \infty, \quad \varepsilon > 0. \tag{4}$$

The existence of a large disparity between (3) and (4) suggests that requirement (4) is too strict and can be replaced by a weaker condition. Levin [5] undertook the first step in this di-

*More rigorous conditions for the finiteness of the number of eigenvalues in the self-adjoint case are given in [4].

rection. He showed that condition (4) can be replaced by an integral-type condition

$$|q(x)| < \exp(-\gamma(x)),$$

where the function $\gamma(x)$ is such that $x\gamma'(x) \to \infty$ as $x \to \infty$ and $\int^{\infty} \gamma(x)\, x^{-2}\, dx = \infty$. Comparing Levin's condition with condition (4), we should note that Levin's condition in fact reduces to a requirement of a linear or almost linear growth of the function $-\ln |q(x)|$ at infinity. The following question then arises. Is it possible to weaken condition (4) appreciably, and to what extent? The answer to this question is given in the first section of the present paper and can be stated as follows: the number of eigenvalues of the operator l_h is finite provided that q(x) satisfies the condition

$$\sup_x |q(x)| \exp(\varepsilon \sqrt{x}) < \infty \tag{5}$$

with some $\varepsilon > 0$. Condition (5) is exact in the sense that the condition

$$\sup |q(x)| \exp(\varepsilon x^{\beta}) < \infty$$

does not ensure the finiteness of the number of eigenvalues for any $\beta \in (0, \frac{1}{2})$ (see [6]). The proof of the last assertion, as well as some other analogous results, will be given in a subsequent paper.

The second section of the present paper is devoted to proving the theorem of expansion in eigenfunctions of the operator l_h.

As is known, expansion in eigenfunctions of nonself-adjoint operators is made difficult by the fact that the resolvent of a nonself-adjoint operator grows rapidly as a spectrum point is approached. A method for overcoming these difficulties based on an idea of Golubev [7] is presented in the present paper. The expansion theorem obtained by us can be considered as an extension of the corresponding results obtained by Naimark.

Finally, the third section is devoted to the generalization of the results of § 1 to the three-dimensional case. The Schroedinger operator with a decreasing complex potential q(x), $x = (x_1, x_2, x_3)$,

$$lu = -\Delta u + q(x)u \tag{6}$$

in three-dimensional space has been studied by Gel'fand [8] and Martirosyan [9]. In particular, it has been shown in [9] that if for some $\varepsilon > 0$

$$\sup_x |q(x)| \exp(\varepsilon |x|) < \infty \tag{7}$$

then the number of eigenvalues of operator (6) is finite. We were able to replace this condition by a weaker one

$$\sup_x |q(x)| \exp(\varepsilon \sqrt{|x|}) < \infty. \tag{8}$$

Let us note in conclusion that all of the above-mentioned results obtained in the present paper are consequences of more general theorems relating the structure of the eigenvalue set to the rate of decrease of q(x) at infinity derived by us. A preliminary discussion of the results of the present paper is contained in a brief note published earlier [6].

§ 1. The One-Dimensional Schroedinger Operator

Several theorems on the structure of the spectrum of a nonself-adjoint differential operator over $[0, \infty)$ with a decreasing potential will be proved in the present section.

Let us consider the following differential expression in the interval $[0, \infty)$:

$$ly = -y'' + q(x)y. \tag{1.1}$$

Here and in the following, unless stated otherwise, q(x) is a continuous complex function satisfying the condition

$$\int_0^\infty x|q(x)|\,dx < \infty. \tag{1.2}$$

We will study the operator l_h obtained by the closure in $L_2(0, \infty)$ of the differential operator defined by expression (1.1) for sufficiently smooth finite functions satisfying the condition

$$y'(0) - hy(0) = 0 \tag{1.3}$$

with h complex. In view of condition (1.2), the operator l_h can be considered as the result of a perturbation applied to the simplest second-order self-adjoint differential operator

$$ly = -y'', \quad y(0) = 0. \tag{1.4}$$

This approach to the study of the operator l_h has been adopted in [1, 2, 3, 5]. It was found that the spectrum of the operator l_h differs little from that of the operator (1.4). The continuous spectrum, as before, fills the half-axis $[0, \infty)$, while the discrete spectrum consists of eigenvalues of finite rank. The set of eigenvalues is not more than denumerable and can only accumulate to the continuous spectrum. On the other hand, if the stronger condition

$$\sup_{0 \leqslant x < \infty} |q(x)| \exp(\varepsilon x) < \infty, \quad \varepsilon > 0 \tag{1.5}$$

is satisfied, then the number of eigenvalues is finite.

There is a considerable disparity between condition (1.5) and (1.2), ensuring the finiteness of the number of eigenvalues in the self-adjoint case. As has already been stated in the Introduction, it is not possible to eliminate this disparity completely. We will show that condition (1.5) can be relaxed. We will obtain the corresponding result as a consequence of more general propositions concerning the behavior of sets of eigenvalue accumulation points as a function of the rate of decrease of the potential.

Let us first of all present the information that will be required below. It is known* that with condition (1.2) satisfied, the equation

$$-y'' + q(x)y = k^2 y, \quad k = \sigma + i\tau, \ \tau \geqslant 0 \tag{1.6}$$

has the unique solution $f(x, k)$ satisfying the condition $f(x, k)\exp(-ikx) \to 1$ as $x \to \infty$. The function $f(x, k)$ satisfies an integral equation of the Volterra type

$$f(x, k) = \exp(ikx) + \int_x^\infty k^{-1} \sin k(t - x)\, q(t) f(t, k)\, dt. \tag{1.7}$$

*See [10, 11]. In these papers it was assumed that $\operatorname{Im} q(x) \equiv 0$, but it is easy to show that the propositions are valid in the case of q(x) complex.

Solving this equation by the method of successive approximations (see [12]), we can show that $f(x, k)$ is a regular function in the half-plane $\tau > 0$ and that the following estimates are valid:

$$|f(x, k) - e^{ikx}| \leqslant K |k|^{-1} \exp(-\tau x) \int_x^\infty |q(t)| dt, \quad \tau \geqslant 0, \tag{1.8}$$

$$|f(x, k) - e^{ikx}| \leqslant K \exp(-\tau x) \int_x^\infty t |q(t)| dt, \quad \tau \geqslant 0, \tag{1.9}$$

$$|f_x(x, k) - ike^{ikx}| \leqslant K \exp(-\tau x) \int_x^\infty |q(t)| dt, \quad \tau \geqslant 0. \tag{1.10}$$

Here and in the following, we use K to denote any constant whose exact value is of no interest to us. It should be noted that estimate (1.8) is useful for large values of k and estimate (1.9) for small values of k.

The function $f(x, k)$ is related to e^{ikx} by means of the transformation operator (see [5])

$$f(x, k) = e^{ikx} + \int_x^\infty A(x, y) e^{iky} dy, \quad \tau \geqslant 0. \tag{1.11}$$

Substituting (1.11) into (1.7), we obtain for the kernel $A(x, y)$ of the transformation operator an equation of the Volterra type

$$A(x, y) = \frac{1}{2} \int_{\frac{x+y}{2}}^\infty q(t) dt - \int_{\frac{x+y}{2}}^\infty dt \int_0^{\frac{y-x}{2}} dz q(t-z) A(t-z, t+z), \quad y \geqslant x.$$

Solving this equation by the method of successive approximations, we obtain the estimates (see [12])

$$|A(x, y)| \leqslant \frac{1}{2} \int_{\frac{x+y}{2}}^\infty |q(t)| dt \exp\left(\int_x^\infty t |q(x)| dt\right), \tag{1.12}$$

$$\left| \frac{\partial}{\partial x} A(x, y) + \frac{1}{4} q\left(\frac{x+y}{2}\right) \right| \leqslant K \int_x^\infty |q(t)| dt \int_{\frac{x+y}{2}}^\infty |q(t)| dt. \tag{1.13}$$

An estimate analogous to (1.13) is also valid for $(\partial/\partial y) A(x, y)$.

In the following we will require additional information about operators of the type of l_h with real potential $q(x)$ satisfying condition (1.2).

Let us consider the following equation containing a complex parameter λ:

$$-y'' + q(x) y = \lambda y. \tag{1.14}$$

Let us use $\varphi_h(x, \lambda)$ and $\psi_h(x, \lambda)$ to denote the solutions of Eq. (1.14) satisfying the boundary conditions

$$\left.\begin{aligned}\varphi_h(0,\lambda)&=\frac{1}{\sqrt{1+h^2}}, & \varphi_h'(0,\lambda)&=\frac{h}{\sqrt{1+h^2}},\\ \psi_h(0,\lambda)&=-\frac{h}{\sqrt{1+h^2}}, & \psi_h'(0,\lambda)&=\frac{1}{\sqrt{1+h^2}},\end{aligned}\right\}\tag{1.15}$$

where h is a real number or $+\infty$. The general solution of Eq. (1.14) is of the form

$$\psi_h(x,\lambda)+m\varphi_h(x,\lambda).$$

Let us determine the number $m = m_h(\lambda)$ from the condition

$$\psi_h(x,\lambda)+m_h(\lambda)\varphi_h(x,\lambda)=\chi(x,\lambda)\in L_2(0,\infty).$$

This can be done uniquely for any imaginary λ, since, with condition (1.2) satisfied we are dealing with the limit-point case.

The function $m_h(\lambda)$ was first studied by Weyl [13]. In the following, we will call it the Weyl function of the operator l_h. The Weyl function is a regular function of the variable λ in each of the half-planes $\operatorname{Im}\lambda > 0$, $\operatorname{Im}\lambda < 0$, and satisfies the condition

$$\operatorname{Im} m_h(\lambda)\operatorname{Im}\lambda<0.\tag{1.16}$$

The spectral function of a self-adjoint operator l_h* can be expressed in terms of $m_h(\lambda)$ as follows:

$$\rho_h(\lambda)=-\lim_{\delta\to 0}\int_0^\lambda \operatorname{Im} m_h(u+i\delta)\,du.\tag{1.17}$$

Weyl functions corresponding to various boundary conditions are related by the expression†

$$\frac{-h_1+m_{h_1}(\lambda)}{h_1 m_{h_1}(\lambda)+1}=\frac{-h_2+m_{h_2}(\lambda)}{h_2 m_{h_2}(\lambda)+1}.\tag{1.18}$$

If $f(x,k)$ is the solution of Eq. (1.14) with $\lambda = k^2$, $\operatorname{Im} k \geq 0$, satisfying the condition $f(x,k)\cdot\exp(-ikx)\to 1$ as $k\to\infty$, then we have

$$m_h(\lambda)=\frac{f(0,k)+hf_x(0,k)}{f_x(0,k)-hf(0,k)}.\tag{1.19}$$

Here and in the following, $k=\sqrt{\lambda}$, where the branch of the square root is chosen such that the λ plane with a branch cut along the real positive axis corresponds to the half-plane $\operatorname{Im} k > 0$.

A knowledge of the Weyl function allows us to give a complete description of the structure of the spectrum of the operator l_h. Namely, with condition (1.2) satisfied, the Weyl function is meromorphic in the plane with a branch cut along the positive real axis filled by the continuous spectrum, while its poles coincide with the eigenvalues.

---

*We use the term "spectral function of the operator l_h" for the nondecreasing function $\rho_h(\lambda)$, such that the expansion in eigenfunctions is of the form

$$f(x)=\frac{1}{\pi}\int_{-\infty}^{\infty}\varphi_h(x,\lambda)\,d\rho_h(\lambda)\int_0^{\infty}\varphi_h(x',\lambda)f(x')\,dx'.$$

Here, $\varphi_h(x,\lambda)$ is the solution of Eq. (1.14) satisfying conditions (1.15).

†In general, formulas (1.17) and (1.18) are always valid when we have the Weyl limit-point case.

With the help of formula (1.18), it is not difficult to determine the Weyl function for complex boundary conditions of the type of (1.3) with $\operatorname{Im} h \neq 0$, $h = \pm i$.* When h is complex, the Weyl function no longer possesses properties (1.16) and (1.17), although the connection between the Weyl function and the spectrum of the operator l_h indicated above remains in force. It can be seen from formula (1.18) that for the description of the spectrum of the operator l_h for any boundary condition of the form (1.3) it is sufficient to know the Weyl function for any one boundary condition, for example,

$$y(0)=0 \quad (h=\infty).$$

Let us return to the investigation of a nonself-adjoint operator l_h with a complex potential.

We will take the complex function q(x) to be of class S_n, n = 0, 1, 2, ..., if it is continuous and $\int_0^\infty |q(x)|x^{n+1}dx<\infty$; we have $q(x)\in S_\infty$ if $q(x)\in S_n$ for any n.

The following simple lemma will be of great importance in the discussions that follow.

Lemma 1.1

If $q(x)\in S_n$ and

$$\int_0^\infty |q(x)|x^{r+1}dx \leqslant C_r < \infty, \quad r=0,\ 1,\ 2,\ \ldots\ n,$$

then the function

$$D_h(k) \equiv f_x(0,\ k) - hf(0,\ k)$$

is regular in the upper half-plane $\operatorname{Im} k > 0$ and continuous together with its first n derivatives for $\operatorname{Im} k \geq 0$; the following estimates are valid for certain $b > 0$:

$$\left|\frac{d^r}{dk^r} D_h(k)\right| \leqslant 2^r\left[\frac{1}{2}\, C_{r-1} + \frac{b}{r+1}\, C_r\right], \qquad r \geqslant 2, \quad \operatorname{Im} k \geqslant 0,$$

$$\sup_{\operatorname{Im} k\geqslant 0}\left|\frac{dD_h(k)}{dk}\right| < \infty, \quad \sup_{\operatorname{Im} k\geqslant 0} |D_h(k) - ik| < \infty.$$

Proof. The regularity of $D_h(k)$ for $\operatorname{Im} k > 0$ follows directly from the properties of the function $f(x, k)$ listed above.

Let us estimate the derivatives of the function $D_h(k)$. In view of (1.18) and (1.19), we obtain for $\operatorname{Im} k \geq 0$

$$|f(0,\ k)| \leqslant 1 + \left|\int_0^\infty A(0,\ y)\, e^{iky}\, dy\right| \leqslant$$
$$\leqslant 1 + \frac{1}{2}\int_0^\infty dy \int_{\frac{y}{2}}^\infty |q(t)|\, dt \exp\left\{\int_0^\infty |q(t)|\, tdt\right\} \leqslant 1 + C_0 e^{C_0}, \tag{1.20}$$

*When h = ±i, the solutions (1.15) become linearly dependent and, therefore, another system of solutions of Eq. (1.14) should be chosen.

$$\left|\frac{d^r f(0,k)}{dk}\right| = \left|\int_0^\infty A(0,y)e^{iky}y^r dy\right| \leqslant \frac{1}{2}e^{C_0}\int_0^\infty y^r \int_{\frac{y}{2}}^\infty |q(t)|\,dt\,dy \leqslant$$
$$\leqslant \frac{2^r}{r+1}e^{C_0}C_r, \quad r \geqslant 1,$$

$$|f_x(0,k) - ik| = \left|\int_0^\infty A_x(0,y)e^{iky}dy\right| \leqslant \frac{1}{4}\int_0^\infty \left|q\left(\frac{y}{2}\right)\right| dy +$$
$$+ K\int_0^\infty |q(t)|\,dt \int_0^\infty dy \int_{\frac{y}{2}}^\infty |q(t)|\,dt \leqslant \int_0^\infty |q(t)|\,dt\left[\frac{1}{2} + 2KC_0\right] =$$
$$= C_{-1}\left[\frac{1}{2} + 2KC_0\right], \tag{1.20}$$

$$\left|\frac{d}{dk}f_x(0,k)\right| \leqslant 1 + \left|\int_0^\infty A_x(0,y)e^{iky}y\,dy\right| \leqslant 1 + 2\left[\frac{1}{2}C_0 + KC_1C_{-1}\right],$$

$$\left|\frac{d^r}{dk^r}f_x(0,k)\right| \leqslant 2^r\left[\frac{1}{2}C_{r-1} + \frac{2}{r+1}KC_rC_{-1}\right], \quad r \geqslant 2, \tag{1.21}$$

where

$$C_{-1} \equiv \int_0^\infty |q(t)|\,dt;$$

from this we find that

$$\left|\frac{d^r D_h(k)}{dk^r}\right| \leqslant 2^r\left[\frac{1}{2}C_{r-1} + \frac{1}{r+1}C_r(2C_{-1}K + |h|e^{C_0})\right], \quad r \geqslant 2.$$

Introducing the abbreviation $2C_{-1}K + |h|e^{C_0} = b$, we obtain the first of the estimates we require; the remaining ones follow directly from (1.20) and (1.21). The lemma has been proved.

We will now establish some propositions concerning the structure of the spectrum of the operator l_h with the help of Lemma 1.1.

We will call the point $\lambda = k^2$ a singular point of the operator l_h if $D_h(k) = 0$.* The multiplicity of the root of $D_h(\sqrt{\lambda})$ if it has a definite value will be called the multiplicity of the singular point λ. The set of all singular points of operator l_h will be denoted by E; the set of all eigenvalues by E_0; the set of all singular points lying on the real positive half-axis $[0, \infty)$ (more accurately, on the upper and lower edges of the branch cut) by E_1; the set of singular points of infinite multiplicity by E_2; the set of all points of accumulation of eigenvalues by E_3. The sets E_0 and E_1 do not intersect and sum to give E. Each singular point belonging to set E_0 has a finite multiplicity equal to the rank of the corresponding eigenvalue. The following theorem holds.

---

* We consider that a point σ_0 on the real axis is a root of the function D(k) if $D(k) \to 0$ as $k \to \sigma_0$ along any nontangential path lying in the upper half-plane. If $q(x) \in S_1$, the function $D_h(k)$ is continuous right up to the real axis (see Lemma 1.1) and its real roots are defined in the conventional manner.

Theorem 1.1

Let $q(x) \in S_1$. Then:

1) the set of eigenvalues numbered with multiplicity taken into account satisfies the condition

$$\sum \operatorname{Im} \sqrt{\lambda_\nu} < \infty;$$

2) $E_3 \subset E_1$;

3) the set E_1 is bounded, closed, has measure zero, and satisfies the condition

$$\sum \ln |l_\nu| |l_\nu| > -\infty, \tag{1.22}$$

where $|l_\nu|$ is the interval in which l_ν is contiguous to the set E_1 and the summation extends over all bounded contiguous intervals.

Proof. If $q(x) \in S_1$, then the function $D_h(k)$ is regular in the half-plane $\operatorname{Im} k > 0$ and continous right up to the real axis. Using formula (1.8) and Lemma 1.1, we can write

$$f(0, k) = 1 + 0(1), \ |k| \to \infty, \ \operatorname{Im} k \geqslant 0,$$

$$D_h(k) = ik + 0(1), \ |k| \to \infty, \ \operatorname{Im} k \geqslant 0.$$

It can be seen from this that all roots of the equation $D_h(k) = 0$ are situated in the finite part of the half-plane $\operatorname{Im} k \geq 0$. From the analyticity of $D_h(k)$ in the half-plane $\operatorname{Im} k > 0$, it follows that the points of the accumulation of roots can only lie on the real axis. Using the continuity of $D_h(k)$ right up to the real axis, we find that $E_1 = \overline{E}_1 E_3 \subset E_1$. From the uniqueness theorem for analytical functions (see [14]), we find that $mE_1 = 0$, while the set of all imaginary roots k_s of equation $D_h(k) = 0$ numbered with due account of multiplicity satisfies the condition

$$\sum \operatorname{Im} k_s < \infty.$$

Finally, condition (1.22) follows from the following result of Beurling (see [15]).

Let the function $g(z)$ be regular in the unit circle, continuous right up to the boundary, and satisfy the Hölder conditions on the boundary. Then, $g(z) \equiv 0$ if $g(e^{i\theta}) = 0$ on the zero-measure set $F \subset [0, 2\pi)$ satisfying the condition

$$\sum |\Delta_\nu| \ln |\Delta_\nu| = -\infty,$$

where $|\Delta_\nu|$ is the length of the interval Δ_ν of contiguity with the set F and the summation extends over all intervals Δ_ν.

In order to apply this result to our case, let us first of all note that the function $(\lambda + i)^{-1} \cdot D_h(\sqrt{\lambda})$ is regular in the half-plane $\operatorname{Im} \lambda > 0$, continuous right up to the real axis, and satisfies the Hölder condition with index $\alpha = \frac{1}{2}$ (because of the change of variable $k = \sqrt{\lambda}$) on the real axis.

Let us show, for example, that the set of real singular points E_1^+ lying on the upper edge of the branch cut $[0, \infty)$ satisfies condition (1.22). For this we map the upper half-plane $\operatorname{Im} \lambda > 0$ into the unit circle. In this process, the function $(\lambda + i)^{-1} D_h(\sqrt{\lambda})$ will change into a function analytic in the unit circle and satisfying the conditions of the Beurling theorem. The required result is obtained when we transform back to the variable λ. Condition (1.22) for the set E_1^- of singular points lying on the lower edge of the branch cut $[0, \infty)$ is verified in exactly the same manner. The theorem has thus been proved.

Let us proceed to the investigation of the spectrum of the operator l_h when $q(x) \in S_\infty$. Applying Lemma 1.1 we find that the function $D_h(k)$ is regular in the half-plane $\operatorname{Im} k > 0$ and continuous right up to the real axis together with all of its derivatives.

We will require the following proposition related to Beurling's theorem given above when we investigate the properties of the function $D_h(k)$.

Lemma 1.2

Let the function g(z) be regular in the unit circle, continuous right up to the boundary together with all of its derivatives, and

$$\sup_{|z|<1} |g^{(r)}(z)| \leqslant A_r, \qquad r = 0, 1, 2, \ldots$$

Then, $g(z) \equiv 0$ when $g(e^{i\theta}) = 0$ together with all derivatives on the zero-measure set $F \subset [0, 2\pi)$ satisfying the following condition:

$$\int_0 \ln T(s)\, d\varphi(F, s) = -\infty, \tag{1.23}$$

where

$$T(s) = \inf_r \frac{A_r s^r}{r!}, \quad r = 0, 1, 2, \ldots,$$

and $\varphi(F, s)$ is the measure of the s-vicinity of the set F.

Proof. Let us assume that $g(z) \not\equiv 0$. Then, by Jensen's formula

$$\int_0^{2\pi} \ln |g(e^{i\theta})|\, d\theta > -\infty. \tag{1.24}$$

On the other hand, using the estimate

$$|g(e^{i\theta})| \leqslant \inf_{\theta_0 \in F} \inf_r \frac{A_r |\theta - \theta_0|^r}{r!},$$

we find that

$$\ln |g(e^{i\theta})| \leqslant \ln T(\inf_{\theta_0 \in F} |\theta - \theta_0|).$$

Next, denoting by F_s the s-vicinity of the set F, we obtain

$$\int_0^{2\pi} \ln |g(e^{i\theta})|\, d\theta = \int_{F_s} \ln |g(e^{i\theta})|\, d\theta + \int_{[0, 2\pi)/F_s} \ln |g(e^{i\theta})|\, d\theta \leqslant$$
$$\leqslant \int_0^s \ln T(s)\, d\varphi(F, s) + \int_{[0, 2\pi)/F_s} \ln |g(e^{i\theta})|\, d\theta.$$

The last integral is bounded since $g(e^{i\theta})$ over the set $[0, 2\pi)/F_S$ has at most a finite number of finitely multiple zeros. Using condition (1.23), we obtain

$$\int_0^{2\pi} \ln |g(e^{i\theta})| = -\infty,$$

so that $g(z) \equiv 0$, $|z| < 1$. The lemma has been proved.

Note. Let the function g(z) be analytic in the half-plane $\operatorname{Im} k > 0$, continuous right up to the real axis together with all derivatives, and let the following conditions be satisfied for some $N > 0$:

$$|g^{(r)}(k)| \leqslant A_r, \quad r = 0, 1, 2, \ldots, \quad \operatorname{Im} k > 0, \ |k| < 2N, \tag{1.25}$$

$$\left| \int_N^{\infty} \frac{\ln |g(\sigma)|}{1+\sigma^2} d\sigma \right| < \infty, \quad \left| \int_N^{\infty} \frac{\ln |g(-\sigma)|}{1+\sigma^2} d\sigma \right| < \infty. \tag{1.26}$$

The function g(k) is identically zero if it becomes zero together with all of its derivatives over a zero-measure set of the real axis satisfying the condition

$$\int_0 \ln T(s) \, d\varphi(F, s) = -\infty.$$

Here, T(s) and $\varphi(F, s)$ have the same meaning as in Lemma 1.2.

The proof of the last proposition differs from that of Lemma 1.2 only in that, instead of inequalities (1.24) we should use the corresponding inequality for the half-plane case (for example, see [14])

$$\int_{-\infty}^{\infty} \ln |g(\sigma)| \frac{d\sigma}{1+\sigma^2} > -\infty.$$

Lemma 1.2 allows us to establish the connection between the rate of decrease of the potential q(x) and the "density" of the set E_3. Let $\widetilde{E}_2$ denote the image in the plane $k = \sqrt{\lambda}$ of the set E_2 of singular points of infinite multiplicity and by $\widetilde{E}_3$, the image of the set E_3. The following theorem describes the structure of the sets $\widetilde{E}_2$ and $\widetilde{E}_3$.

Theorem 1.2

Let $q(x) \in S_\infty$. Then the set $\widetilde{E}_2$ is bounded, closed, is of measure zero, and $\widetilde{E}_3 \subset \widetilde{E}_2$. Moreover, the following condition is satisfied:

$$\int_0 \ln T_l(s) \, d\varphi(\widetilde{E}_2, s) > -\infty, \tag{1.27}$$

where

$$T_l(s) = \inf_r \left[\frac{1}{2} C_{r-1} + \frac{b}{r+1} C_r \right] \frac{(2s)^r}{r!},$$

C_r and b are as defined in Lemma 1.1, and $\varphi(\widetilde{E}_2, s)$ is the linear measure of the s-vicinity of the set $\widetilde{E}_2$.

Proof. Only the last proposition requires proof. Let us choose N to be so large that the function $D_h(k)$ has no roots in the intervals $(-\infty, -N)$, (N, ∞). In view of inequality (1.20), we have

$$\left|\int_{N+1}^{\infty} \frac{\ln|D_h(\sigma)|}{1+\sigma^2}\,d\sigma\right| < \infty, \quad \left|\int_{N+1}^{\infty} \frac{\ln|D_h(-\sigma)|}{1+\sigma^2}\,d\sigma\right| < \infty.$$

Relying on the remark to Lemma 1.2 and using the estimate (see Lemma 1.1)

$$|D_h^{(r)}(k)| \leqslant 2^r\left[\frac{1}{2}C_{r-1} + \frac{b}{r+1}C_r\right], \quad r > 2,$$

we obtain the required result.

Corollary 1. If the potential q(x) satisfies the condition

$$|q(x)| \leqslant C\exp(-\delta x^\alpha), \quad \delta > 0, \quad 0 < \alpha < \frac{1}{2}, \tag{1.28}$$

then the set $\tilde{E}_2$ is bounded, closed, is of linear measure zero, and satisfies the condition

$$\sum |l_\nu|^{\frac{1-2\alpha}{1-\alpha}} < \infty. \tag{1.29}$$

Here, $|l_\nu|$ is the length of the interval l_ν of contiguity with the set $\tilde{E}_2$ and the summation extends over all bounded intervals l_ν.

Indeed, in the present case, $q(x) \in S_\infty$ and, consequently, the conditions of Theorem 1.2 are fulfilled; moreover,

$$C_r = \int_0^\infty x^{r+1}|q(x)|\,dx \leqslant C\int_0^\infty x^{r+1}\exp(-\delta x^\alpha)\,dx < Bd^{r+1}(r+1)^{\frac{r+1}{\alpha}},$$

where

$$B = C\int_0^\infty \exp\left(-\frac{\delta}{2}x^\alpha\right)dx, \quad d = \left(\frac{2}{e\delta\alpha}\right)^{\frac{1}{\alpha}}.$$

In view of Lemma 1.1, the derivatives of the function $D_h(k)$ have the estimates

$$\left|\frac{d^r D_h(k)}{dk^r}\right| \leqslant 2^r\left[\frac{1}{2}Bd^r r^{\frac{r}{\alpha}} + B\frac{b}{r+1}d^{r+1}(r+1)^{\frac{r+1}{\alpha}}\right].$$

Using the inequalities

$$\left(1+\frac{1}{r}\right)^{\frac{r}{\alpha}} < e^{\frac{1}{\alpha}}, \quad (r+1)^{\frac{1}{\alpha}-1} < \exp\left(\frac{r+1}{\alpha}\right),$$

we can write

$$\left|\frac{d^r D_h(k)}{dk^r}\right| \leqslant d_1^r B_1 r^{\frac{r}{\alpha}},$$

where

$$d_1 = 2de^{\frac{1}{\alpha}}, \quad B_1 = Be^{\frac{1}{\alpha}}\left[1 + dbe^{\frac{1}{\alpha}}\right].$$

Applying the inequality $r^r < r!e^r$, we obtain

$$T_l(s) \leqslant \inf_r \frac{B_1 d_1 r^{\frac{r}{\alpha}}}{r!} s^r \leqslant B_1 \inf_r r^{\left(\frac{1}{\alpha}-1\right)r} (ed_1 s)^r \leqslant B_1 \exp\left\{-\frac{1-\alpha}{\alpha} e^{-\frac{1}{1-\alpha}} d_1^{-\frac{\alpha}{1-\alpha}} \frac{1}{s^{\frac{\alpha}{1-\alpha}}}\right\}.$$

It follows from this and (1.27) that

$$\int_0^{2l_0} \frac{d\varphi(\widetilde{E}_2, s)}{s^{\frac{\alpha}{1-\alpha}}} < \infty. \tag{1.30}$$

Here, l_0 denotes the length of the longest of the contiguity intervals.

Let ψ (s) denote the sum of the length of the intervals of contiguity to the set $\widetilde{E}_2$, the lengths of which do not exceed s, i.e.,

$$\psi(s) = \sum_{|l_\nu| \leqslant s} |l_\nu|.$$

It is not difficult to see that the measure $\varphi(\widetilde{E}_2, s)$ of the s-vicinity of set E_2 can be expressed in terms of ψ (s) as

$$\varphi(\widetilde{E}_2, s) = 2s \int_{2s}^{2l_0} \frac{d\psi(t)}{t} + \int_0^{2s} d\psi(t) + 2s. \tag{1.31}$$

It should be noted that the first integral on the right-hand side of (1.31) gives the number of finite contiguity intervals whose lengths exceed s. It follows from (1.30) and (1.31) that

$$\int_\varepsilon^{2l_0} s^{-\frac{\alpha}{1-\alpha}} d\varphi(\widetilde{E}_2, s) = 2\int_\varepsilon^{2l_0} s^{-\frac{\alpha}{1-\alpha}} \int_{2s}^{2l_0} \frac{d\psi(t)}{t} ds + 2\int_\varepsilon^{2l_0} s^{-\frac{\alpha}{1-\alpha}} ds < \int_0^{2l_0} s^{-\frac{\alpha}{1-\alpha}} d\varphi(\widetilde{E}_2, s) < \infty.$$

Proceeding to the limit as $\varepsilon \to 0$, we obtain, with $\alpha \in (0, 1/2)$

$$2\int_0^{2l_0} s^{-\frac{\alpha}{1-\alpha}} \int_{2s}^{2l_0} \frac{d\psi(t)}{t} ds < \infty.$$

Let $n(|l|)$ denote the number of intervals of contiguity to $\widetilde{E}_2$ whose lengths are all equal to $|l|$. Assuming that the contiguity intervals are numbered in order of decreasing length, we obtain

$$2\int_0^{2l_0} s^{-\frac{\alpha}{1-\alpha}} \int_{2s}^{2l_0} \frac{d\psi(t)}{t} ds = 2^{\frac{\alpha}{1-\alpha}} \sum_\nu \int_{|l_{\nu+1}|}^{|l_\nu|} s^{-\frac{\alpha}{1-\alpha}} ds \sum_{|l| \geqslant |l_\nu|} n(|l|) = 2^{\frac{\alpha}{1-\alpha}} \frac{1-\alpha}{1-2\alpha} \sum_\nu |l_\nu|^{\frac{1-2\alpha}{1-\alpha}}.$$

The validity of condition (1.29) has been proved.

Corollary 2. If

$$\int\limits_0 \ln T_l(s)\,ds = -\infty, \tag{1.32}$$

then the set $\tilde{E}_2$ is a null set and, consequently, the total number of singular points of the operator l_h is finite for any h, while all singularities are of finite multiplicity. In particular, the number of eigenvalues of l_h is finite.

Condition (1.32) is satisfied, for example, if for any $C > 0$, $\delta > 0$,

$$|q(x)| < C\exp(-\delta\sqrt{x}). \tag{1.33}$$

Indeed, it follows from (1.30) that the set $\tilde{E}_2$ in this case must satisfy the condition

$$\int\limits_0 \frac{d\varphi(\tilde{E}_2, s)}{s} < \infty,$$

while this can be satisfied if, and only if, $\varphi(\tilde{E}_2, s) = 0$ for all s, i.e., $\tilde{E}_2$ is a null set.

It is interesting to compare the above condition for the finiteness of the number of eigenvalues with condition (1.5). As we have already mentioned in the Introduction, condition (1.5) is in fact the condition for the regularity of the function $D_h(k)$ in the half-plane $2\,\mathrm{Im}\,k > -\varepsilon$. This follows, for example, from the representation

$$D_h(k) = ik + \int\limits_0^\infty A_x(0, y)e^{iky}\,dy - h\int\limits_0^\infty A(0, y)e^{iky}\,dy - h$$

and estimates (1.12) and (1.13). The regularity of $D_h(k)$ in the half-plane $2\,\mathrm{Im}\,k > -\varepsilon$ excludes the possibility that eigenvalues of the operator l_h can accumulate to the real axis.

Levin has shown that condition (1.5) can be replaced by the weaker one

$$|q(x)| < \exp(-\gamma(x)), \tag{1.34}$$

where the function $\gamma(x)$ is such that $x\gamma'(x) \to \infty$ as $x \to \infty$ and $\int^\infty \gamma(x)x^{-2}\,dx = \infty$. Condition (1.34) ensures the quasi-linearity of the function $D_h(k)$ on the real axis. It should be recalled in this connection that in the self-adjoint case the number of eigenvalues of the operator l_n is already finite when condition (1.2) is satisfied. The question arises now of whether it is possible to reduce further the disparity between the conditions for the finiteness of the number of eigenvalues in the self-adjoirt and nonself-adjoint cases.

The result [condition (1.33)] obtained by us is an answer to this question. In [6] it was noted that the condition

$$|q(x)| < C\exp(-\delta x^\alpha),\ C > 0,\ \delta > 0$$

does not ensure the finiteness of the number of eigenvalues of the operator l_h for any $\alpha < 1/2$. In this connection, it is of interest to compare condition (1.32) with the condition for the quasi-analyticity of the function $D_h(k)$ on the real axis. According to the Carleman–Ostrovskii theorem (for example, see [16]), in order for the function $D_h(\sigma)$, $\sigma \in (-\infty, \infty)$, to belong to the quasi-analytic class, it is necessary and sufficient that all of its derivatives have the estimates

$$|D_h^{(r)}(\sigma)| \leqslant A_r = CM^r m_r, \qquad r \geqslant 1$$

with numbers m(r) satisfying the condition

$$\int_1^\infty \frac{\ln \Theta_l(\rho)}{\rho^2}\, d\rho = \infty. \tag{1.35}$$

Here, $\Theta_l(\rho) = \sup_r \frac{\rho^r}{m_r}$. Condition (1.35) can be rewritten as

$$\int_0 \ln T_l'(s)\, ds = -\infty,$$

where

$$T_l'(s) = \inf_r A_r s^r.$$

Using the notation of Lemma 1.2, we obtain

$$\int_0 \ln T_l(s)\, ds < \int_0 \ln T_l'(s)\, ds.$$

It can be seen directly from this that any function $D_h(\sigma)$ belonging to the quasi-analytic class satisfies condition (1.32). The reverse is incorrect.

In conclusion, it should be noted that the analyticity of the potential q(x) at infinity allows us to make a more detailed study of the analytical properties of the function $D_h(k)$ and to state stronger assertions about the spectrum. In particular, Lidskii has noted that the regularity of the potential q(x) in the vicinity of an infinitely remote point of the complex plane x, together with a decay of order of x^{-2}, secures the finiteness of the number of eigenvalues of an operator of type (1.1).

In [17]* the potential q(x) is assumed to be regular at infinity in a sector $|\arg x| < \theta$ and to decrease along rays in this sector faster than x^{-2}. It follows from this that the function $D_h(k)$ is regular in the sector

$$\left|\arg k - \frac{\pi}{2}\right| < \frac{\pi}{2} + \theta$$

containing the upper half-plane. It follows from this that the singular points of the operator (1.1) can accumulate only to zero. If we exclude the possibility of accumulation to zero by assuming in addition that, for example, the operator (1.1) is dissipative (see [3]), then the number of singular points is finite.

§ 2. Expansion Theorem

In the present section we give a proof of the theorem of expansion in eigenfunction and adjoint functions of the operator l_h first given in [6]. Our results can be considered as an exten-

* Paper [17] contains an error. Namely, from the analyticity of the potential in sector $|\arg x| < \pi/2$ a conclusion is drawn that the function D(k) is integral. In fact it can have a branch cut along the negative imaginary axis. As an example, we can give the Yukawa potential $q(x) = x^{-1}\exp(-\alpha x)$, well studied by physicists. The cut of D(k) in this case rungs along $(-i\alpha, -i\infty)$.

sion of the corresponding results of Naimark [3]. For simplicity, we will restrict our discussion to the case when the operator l_h has no imaginary singular points. The general case can be reduced to the case being considered in the same way as was indicated in [6].

Let $f_1(x)$ and $f_2(x)$ be sufficiently smooth finite functions, $m_h(\lambda)$ the Weyl function of the operator l_h,* and $\varphi_h(x, \lambda)$ the solutions of the homogeneous equation introduced in § 1 [formulas (1.14) and (1.15)]. Further, let

$$\tilde{f}_k(\lambda) = \int_0^\infty \varphi(x, \lambda) f_k(x)\, dx, \qquad k = 1, 2.$$

As is known, $\tilde{f}_k(\lambda)$ are integral functions and the Parseval equality in complex form (see [18]) holds:

$$\int_0^\infty f_1(x)\bar{f}_2(x)\, dx = \frac{1}{2\pi i}\oint_\gamma \tilde{f}_1(\lambda)\,\tilde{\bar{f}}(\lambda)\, m_h(\lambda)\, d\lambda. \tag{2.1}$$

Here, γ is the contour encompassing the spectrum of the operator l_h. In our case the contour encompasses the positive half-axis $0 \le \lambda < \infty$.

In order to obtain the extended expansion theorem we must transform the right-hand side of formula (2.1) from an integration around the spectrum to an integration over the spectrum. The possibility of such a transformation is based on a theorem of Khavin (see [7]), although it can be effectively applied only in the simplest case when the Weyl function satisfies the following condition near the spectrum of the operator l_h:†

$$|m_h(\lambda)| \leqslant \frac{C}{|\operatorname{Im}\lambda|^p}, \quad 0 \leqslant p < \infty. \tag{2.2}$$

In the present section we will first of all prove the following theorem.

Theorem 2.1

If the Weyl function of the operator l_h satisfies condition (2.2), then the Parseval formula can be written as

$$\int_0^\infty f_1(x)\bar{f}_2(x)\, dx = \int_0^N \rho(\lambda)\left\{\left(\frac{d}{d\lambda}\right)^{p+1}\tilde{f}_1(\lambda)\tilde{\bar{f}}_2(\lambda)\right\} d\lambda + \int_0^\infty \sigma(\lambda)\tilde{f}_1(\lambda)\tilde{\bar{f}}_2(\lambda)\, d\lambda. \tag{2.3}$$

Here, $0 < N < \infty$, while $\sigma(\lambda)$ and $\rho(\lambda)$ are piece-wise continuous complex functions.

Proof. In view of the theorems of § 1, the set of singular points of operator l_h is bounded when condition (1.2) is satisfied. Therefore, we can choose a number N such that in the interval $[N, \infty)$ there are no singular points of the operator l_h. This allows us to represent the Weyl function as the sum of two functions

* Here, as before, l_h is the operator generated in $L_2(0, \infty)$ by the differential expression (1.1) and boundary condition (1.3). The potential $q(x)$ of the operator l_h satisfies condition (1.2).

† Condition (2.2) is satisfied, for example, for operators with rapidly decreasing potentials

$$|q(x)| \leqslant C\exp(-\delta\sqrt{x}).$$

$$m_h(\lambda) = m^N(\lambda) + m^\infty(\lambda),$$

the first of which is regular at infinity and has all singularities in the interval [0, N], while the second is regular in the λ plane with a branch cut $[N, \infty)$ and is continuous right up to the cut. In accordance with this, the contour integral appearing on the right-hand side of formula (2.1) can be subdivided into two integrals

$$\frac{1}{2\pi i}\oint_\gamma m_h(\lambda)\tilde{f}_1(\lambda)\tilde{\tilde{f}}_2(\lambda)\,d\lambda = \frac{1}{2\pi i}\oint_\gamma m^N(\lambda)\tilde{f}_1(\lambda)\tilde{\tilde{f}}_2(\lambda)\,d\lambda + \frac{1}{2\pi i}\oint_\gamma m^\infty(\lambda)\tilde{f}_1(\lambda)\tilde{\tilde{f}}_2(\lambda)\,d\lambda \equiv I_N + I_\infty.$$

It should be noted that the integration contour in the first integral can be closed, while in the second integral the contour can be deformed to the real axis because of the continuity of $m^\infty(\lambda)$

$$I_N = \frac{1}{2\pi i}\oint_{\gamma_N} m^N(\lambda)\tilde{f}_1(\lambda)\tilde{\tilde{f}}_2(\lambda)\,d\lambda,$$

$$I_\infty = \int_N^\infty \frac{-m^\infty(\lambda + i0) + m^\infty(\lambda - i0)}{2\pi i}\tilde{f}_1(\lambda)\tilde{\tilde{f}}_2(\lambda)\,d\lambda.$$

The second integral is already in the required form, while the first requires some further transformations. To do this, we represent $m^N(\lambda)$ as the sum of two functions

$$m^N(\lambda) = M(\lambda) + N(\lambda),$$

the first of which is continuous right up to the real axis and is regular at infinity, while the second has a zero at infinity of order $p + 2$ {all singularities of the functions $M(\lambda)$ and $N(\lambda)$ lie on the segment [0, N]}. This can be achieved, for example, in the following manner. Let $C_1, C_2, \ldots, C_{p+1}$ denote the first $p + 1$ coefficients in the expansion of $m^N(\lambda)$ in a Taylor series in $1/\lambda$. Let us solve the truncated moment problem

$$\int_0^N \tau(t)\,dt = C_1, \qquad \int_0^N t\tau(t)\,dt = C_2,$$

$$\int_0^N t^p\tau(t)\,dt = C_{p+1}.$$

If we seek the solution $\tau(t)$ in the form of a polynomial

$$\tau(t) = a_0 + a_1 t + a_2 t^2 + \ldots + a_p t^p,$$

then for the determination of the coefficients we obtain a system of linear equations whose determinant is different from zero (see [19]). As $M(\lambda)$ we can now take the function

$$M(\lambda) = \int_0^N \frac{\tau(t)}{\lambda - t}\,dt, \tag{2.4}$$

while for $N(\lambda)$ we can take

$$m^N(\lambda) - \int_0^N \frac{\tau(t)}{\lambda - t}\,dt.$$

Indeed, it is not difficult to establish that the expansion in powers of $1/\lambda$ of the function $N(\lambda)$ defined in this way begins at least with a term of form C/λ^{p+2}. Indeed, we have

$$m^N(\lambda)-\int_0^N \frac{\tau(t)}{\lambda-t}dt=\frac{C_1}{\lambda}+\frac{C_2}{\lambda^2}+\ldots+\frac{C_{p+1}}{\lambda^{p+1}}+\frac{C_{p+2}}{\lambda^{p+2}}+\ldots$$
$$+\int_0^N \tau(t)\,dt\frac{1}{\lambda}-\int_0^N \tau(t)\,t\,dt\frac{1}{\lambda^2}-\ldots-\int_0^N \tau(t)\,t^p dt\frac{1}{\lambda^{p+1}}-$$
$$-\int_0^N \tau(t)\,t^{p+1}dt\frac{1}{\lambda^{p+2}}-\ldots=\frac{1}{\lambda^{p+2}}\left(C_{p+2}-\int_0^N \tau(t)\,t^{p+1}dt\right)+$$
$$+\frac{1}{\lambda^{p+3}}\left(C_{p+3}-\int_0^N \tau(t)\,t^{p+2}dt\right)+\ldots$$

The presence of other required properties of the functions $M(\lambda)$ and $N(\lambda)$ is obvious.

Finally, let us introduce the function

$$\mu(\lambda)=\int_\infty^\lambda dz_1\int_\infty^{z_1} dz_2\ldots\int_\infty^{z_{p-1}} N(z_p)\,dz_p=\int_\infty^\lambda N(z)\frac{(\lambda-z)^p}{p!}dz.$$

In view of the fact that $N(\lambda)$ possesses a zero of order $p+2$ at infinity, the function $\mu(\lambda)$ is regular and single-valued in the λ plane with cut $[0, N]$. Moreover, in view of condition (2.2), it is bounded and continuous right up to the cut and, consequently, it can be represented by a Cauchy integral

$$\mu(\lambda)=\int_0^N \frac{\rho(t)}{t-\lambda}d\lambda$$

with a continuous density $\rho(t)$.

From what has been said follows the possibility of the representation of the function $N(\lambda)$ as

$$N(\lambda)=\left(\frac{d}{d\lambda}\right)^{p+1}\mu(\lambda)=(p+1)!\int_0^N \frac{\rho(t)}{(t-\lambda)^{p+2}}dt. \tag{2.5}$$

Combining formulas (2.4) and (2.5), we obtain

$$m^N(\lambda)=-\int_0^N \frac{\tau(t)}{t-\lambda}dt+(p+1)!\int_0^N \frac{\rho(t)}{(t-\lambda)^{p+2}}dt.$$

Parseval's equality (2.1) now takes the form

$$\int_0^\infty f_1(x)\bar{f}_2(x)\,dx=\frac{1}{2\pi i}\oint\left[-\int_0^N \frac{\tau(t)}{t-\lambda}dt+(p+1)!\int_0^N \frac{\rho(t)}{(t-\lambda)^{p+2}}dt\right]\times$$

$$\times \tilde{f}_1(\lambda)\tilde{\tilde{f}}_2(\lambda)\,d\lambda + \int_N^\infty \frac{m(\lambda - i0) - m(\lambda + i0)}{2\pi i}\tilde{f}_1(\lambda)\tilde{\tilde{f}}_2(\lambda)\,d\lambda.$$

After a change in the order of integration in the first integral on the right, we obtain Parseval's equality in the form

$$\int_0^\infty f_1(x)\overline{f}_2(x)\,dx = \int_0^N [-\tau(t)]\,\tilde{f}_1(t)\tilde{\tilde{f}}_2(t)\,dt + \int_0^N \rho(t)\left(\frac{d}{dt}\right)^{p+1}\left\{\tilde{f}_1(t)\tilde{\tilde{f}}_2(t)\right\}dt + \int_N^\infty \frac{m(t-i0)-m(t+i0)}{2\pi i}\tilde{f}_1(t)\tilde{\tilde{f}}_2(t)\,dt, \tag{2.6}$$

which, as can be easily seen, coincides with expression (2.3). The theorem has been proved. Now, it is easy to obtain the expansion of arbitrary finite functions in terms of the principal functions of the operator l_h. For this, we introduce the "adjoint functions" over the continuous spectrum defined by

$$\varphi^{(s)}(\lambda, x) \equiv \frac{d^s}{d\lambda^s}\varphi(\lambda, x), \quad \varphi^0(\lambda, x) \equiv \varphi(\lambda, x),$$

and we rewrite (2.6) in the form

$$\int_0^\infty f_1(x)\overline{f}_2(x)\,dx = \int_0^N [-\tau(t)]\int_0^\infty f_1(x)\varphi^0(t, x)\,dx \times \int_0^\infty \overline{f}_2(y)\varphi^0(t, y)\,dy\,dt + \int_N^\infty \frac{m(t-i0)-m(t+i0)}{2\pi i}\,dt\int_0^\infty f_1(x)\varphi^0(t, x)\,dx \times$$

$$\times \int_0^\infty \overline{f}_2(y)\varphi^0(t, y)\,dy + \int_0^N \rho(t)\sum_{s=0}^{p+1} C_{p+1}^s \int_0^\infty f_1(x)\varphi^{(s)}(t, x)\,dx \times \int_0^\infty \overline{f}_2(y)\varphi^{(p+1-s)}(t, y)\,dy\,dt \equiv I_1 + I_2 + I_3.$$

The order of integration with respect to t and y in integrals I_1 and I_3 can be interchanged in view of the finiteness of $f_1(x)$ and $f_2(y)$. We can do the same in integral I_2 by making use, for example, of the smoothness of the functions $f_1(x)$, $f_2(y)$ and the asymptotic behavior of $m_h(\lambda)$ as $\lambda \to \infty$ (see §1). Then, using the arbitrariness in the choice of the function $f_2(y)$, we obtain the expansion of $f_1(x)$ in the form

$$f_1(x) = \int_0^N \varphi^0(t, x)[-\tau(t)]\int_0^\infty \varphi^0(t, x')f_1(x')\,dx'dt + \int_N^\infty \varphi^0(t, x)\frac{m(t-i0)-m(t+i0)}{2\pi i}\int_0^\infty \varphi^0(t, x')f_1(x')\,dx'dt + \sum_{s=0}^{p+1} C_{p+1}^s \int_0^N \varphi^{(s)}(t, x)\rho(t)\int_0^\infty \varphi^{(p+1-s)}(t, x')f(x')\,dx'dt. \tag{2.7}$$

It should be noted that the function $\tau(t)$ and $\rho(t)$ in formula (3.7) are not uniquely defined.

Until now we have assumed that in the vicinity of points of the real axis the Weyl function everywhere satisfies condition (2.2). Let us assume now that the Weyl function obeys a series of conditions

$$|m_h(u+iv)| \leqslant \frac{C_r}{v^{p_r}}, \quad r=1, 2, \ldots, k \cdot$$

for $a_r \leq u \leq b_r$, $0 \leq p_r < \infty$ when the half-axis $[0, \infty)$ is covered by a finite number of segments* $[a_r, b_r]$. Then, the procedure just described in application to the segment $[0, N]$ can be carried out for each segment $[a_r, b_r]$. The formula analogous to (2.7) obtained in this way can be written as

$$f(x)=\int_0^\infty \tau(t)\varphi^0(t, x)\int_0^\infty \varphi^0(t, x')f(x')dx'dt +$$
$$+\sum_{l=1}^{k}\sum_{s_l=0}^{p_l+1} C_{p_l+1}^{s_l}\int_{a_l}^{b_l}\varphi^{(s_l)}(t, x)\rho_l(t)\int_0^\infty \varphi^{(p_l+1-s_l)}(t, x')f(x')dx'dt. \tag{2.8}$$

If at this stage we introduce the concept of "spectral function" of a nonself-adjoint operator in accordance with (1.1), then the result obtained can be formulated as follows.

Theorem 2.2

If on the segment $[a, b]$ the Weyl function of the operator l_h satisfies the condition†

$$|m_h(u+iv)| \leqslant \frac{C}{v^p}, \quad a \leqslant u \leqslant b,$$

then the spectral function of operator l_h in this segment is a generalized function of order not exceeding p.

In particular, as follows from the results of §1 of the present paper, the spectral function of operator l_h with a rapidly decreasing potential q(x)

$$|q(x)| \leqslant C\exp[-\varepsilon\sqrt{x}]$$

is always a generalized function of finite order.

§ 3. The Three-Dimensional Schroedinger Operator

In the present section we will formulate and prove several propositions concerning the spectrum of the Schroedinger operator in three-dimensional space R_3.

Let us consider in $L_2(R_3)$ an operator L of the Schroedinger type obtained by the closure of an operator defined initially over sufficiently smooth finite functions by the differential expression

$$-\Delta u+q(x)u. \tag{3.1}$$

---

* The last segment $[a_k, b_k]$ has been obtained by the addition of the point $b_k = \infty$ to the semi-infinite interval $[a_k, \infty)$. According to the theorems of § 1, we can always take $p_k = 0$.

† This will be the case, for example, if singular points of operator l_h with multiplicity not exceeding p lie in $[a, b]$.

Unless stated otherwise, here and in the following q(x) is a continuous complex function of three variables (x_1, x_2, x_3) satisfying the condition

$$|q(x)| \leqslant \frac{C}{1+|x|^{3+\varepsilon}}, \quad \varepsilon > 0, \quad |x| = \sqrt{x_1^2 + x_2^2 + x_3^2}.$$

The resolvent $R(\lambda)$ of the operator L is an integral operator. The problem of the study of the spectrum of the operator L can be reduced, as is known (see [20]), to the study of the integral equation which is satisfied by the resolvent kernel $R(x, y, \lambda)$,

$$R(x, y, \lambda) = \frac{1}{4\pi} e^{i\sqrt{\lambda}|x-y|} - \frac{1}{4\pi} \int_{R_3} \frac{e^{i\sqrt{\lambda}|x-z|}}{|x-z|} q(z) R(z, y, \lambda)\, dz. \tag{3.2}$$

This integral equation can be easily obtained with the help of Green's formula from the differential equations

$$-\Delta_x R(x, y, \lambda) + q(x) R(x, y, \lambda) = \lambda R(x, y, \lambda) + \delta(x-y),$$

$$-\Delta_x R_0(x, y, \lambda) = \lambda R_0(x, y, \lambda) + \delta(x-y)$$

for the kernel of the resolvent of operator L and for kernel of the resolvent $R_0(x, y, \lambda) = (4\pi)^{-1} \exp(i\sqrt{\lambda}\,|x-y|)$ of the operator L_0 with potential $q_0(x) \equiv 0$.

Integrating Eq. (3.2) once and introducing the abbreviation

$$R(x, y, \lambda) - \frac{1}{4\pi} e^{i\sqrt{\lambda}|x-y|} \equiv v(x, y, \lambda),$$

we obtain the integral equation for the function $v(x, y, \lambda)$,

$$v(x, y, \lambda) = I_0(x, y, \lambda) + \int_{R_3} I(x, z, \lambda)\, q(z)\, v(z, y, \lambda)\, dz, \tag{3.3}$$

where

$$I(x, y, \lambda) = \frac{1}{16\pi^2} \int_{R_3} \frac{e^{i\sqrt{\lambda}|x-z|}}{|x-z|} q(z) \frac{e^{i\sqrt{\lambda}|z-y|}}{|z-y|}\, dz,$$

$$I_0(x, y, z) = -I(x, y, z) + \int_{R_3} I(x, z, \lambda) \frac{e^{i\sqrt{\lambda}|z-y|}}{4\pi|z-y|}\, dz.$$

For all λ satisfying the condition $\operatorname{Im}\sqrt{\lambda} > 0$, the integral operator with kernel $I(x, z, \lambda)q(z)$ is a completely continuous operator in the space $C^0(R_3)$ of continuous functions tending to zero at infinity (see [20]). Consequently, we can apply Fredholm's theory (see [9]) to it and thus the following representation can be obtained for the function $v(x, y, \lambda)$:

$$v(x, y, \lambda) = \frac{1}{D(\sqrt{\lambda})} \int_{R_3} K(x, z, \lambda)\, I_0(z, y, \lambda)\, dz, \tag{3.4}$$

where

$$K(x, z, \lambda) = I_\lambda\begin{pmatrix} x \\ z \end{pmatrix} - \int_{R_3} I_\lambda\begin{pmatrix} x\xi \\ z\xi \end{pmatrix} d\xi + \frac{1}{2!} \int_{R_3}\int_{R_3} I_\lambda\begin{pmatrix} x\xi_1\xi_2 \\ z\xi_1\xi_2 \end{pmatrix} d\xi_1 d\xi_2 + \ldots \tag{3.5}$$

$$\ldots+\frac{(-1)^n}{n!}\int\limits_{R_3}\int\limits_{R_3}\ldots\int\limits_{R_3} I_\lambda\begin{pmatrix}x\xi_1\xi_2 & \ldots & \xi_n\\ z\xi_1\xi_2 & \ldots & \xi_n\end{pmatrix}d\xi_1 d\xi_2\ldots d\xi_n+\ldots, \tag{3.5}$$

$$D(\sqrt{\lambda})=1-\int\limits_{R_3} I_\lambda\begin{pmatrix}\xi\\ \xi\end{pmatrix}d\xi+\frac{1}{2!}\int\limits_{R_3}\int\limits_{R_3} I_\lambda\begin{pmatrix}\xi_1\xi_2\\ \xi_1\xi_2\end{pmatrix}d\xi_1 d\xi_2+\ldots$$
$$\ldots+\frac{(-1)^n}{n!}\int\limits_{R_3}\int\limits_{R_3}\ldots\int\limits_{R_3} I_\lambda\begin{pmatrix}\xi_1\xi_2 & \ldots & \xi_n\\ \xi_1\xi_2 & \ldots & \xi_n\end{pmatrix}d\xi_1 d\xi_2\ldots d\xi_n+\ldots \tag{3.6}$$

In the last two formulas we have used the usual notation

$$I_\lambda\begin{pmatrix}x_1x_2 & \ldots & x_n\\ y_1y_2 & \ldots & y_n\end{pmatrix}=\begin{vmatrix}I(x_1,y_1,\lambda)\,I(x_1,y_2,\lambda)\ldots I(x_1,y_n,\lambda)\\ I(x_2,y_1,\lambda)\,I(x_2,y_2,\lambda)\ldots I(x_2,y_n,\lambda)\\ \cdots\cdots\cdots\cdots\\ I(x_n,y_1,\lambda)\,I(x_n,y_2,\lambda)\ldots I(x_n,y_n,\lambda)\end{vmatrix}\prod_{i=1}^{n} q(y_i).$$

We will show that with certain additional conditions on $q(x)$ series (3.6) converges absolutely, while series (3.5) converges absolutely and uniformly with respect to $x, y \in R_3$ for all λ satisfying condition $\operatorname{Im}\sqrt{\lambda} > 0$. It follows from this that the functions $K(x, z, \lambda)$ and $D(\sqrt{\lambda})$ are regular in the plane λ with a cut along the positive real half-axis. It is not difficult to see that the resolvent R_λ of the operator L will be a bounded integral operator for all values of λ outside the positive real half-axis $[0, \infty)$ that are not roots of the function $D(\sqrt{\lambda})$. Since the operator L has no eigenvalues on the half-axis $\lambda > 0$ (see [21]), we can assert that all eigenvalues of the operator L with the exception, possibly, of $\lambda = 0$, are roots of the equation

$$D(\sqrt{\lambda})=0. \tag{3.7}$$

From the considerations given above, we see that the function $D(\sqrt{\lambda})$ plays the same role in three-dimensional cases as that played by the function $D_h(\sqrt{\lambda})$ in the one-dimensional case. Lemma 1.1 is now replaced by the following lemma.

Lemma 3.1

Let $q(x)$ be a continuous complex function satisfying the condition

$$\sup_x |q(x)|\,(1+|x|^{3+\varepsilon})\,|x|^s\leqslant C_s<\infty,\quad s=0,\,1,\,\ldots,\,2m. \tag{3.8}$$

Then, $D(k)$ is a regular function of k in the half-plane $\operatorname{Im} k > 0$ and for arbitrary $D(k)$ the following estimate, uniform with respect to k, $\operatorname{Im} k \geq 0$, is valid:

$$|D^{(s)}(k)|\leqslant F_0\, 6^{2s} e^s s!\, Q(s),\quad s\leqslant m, \tag{3.9}$$

where

$$Q(s)=\max_{\substack{s_i\geqslant 1\\ \Sigma s_i=s}}\prod_i \frac{C_{s_i}+\sqrt{C_{2s_i}}}{s_i!},\quad s_i=1,\,2,\,3,\,\ldots,$$

while F_0 is a number depending only on C_0.

Proof. Let us note first of all that each term of the series (3.6) is a regular function of k ($k \equiv \sqrt{\lambda}$) in the half-plane $\operatorname{Im} k > 0$. Indeed, $I(x, y, k^2)$ is a regular function of k in the half-plane $\operatorname{Im} k > 0$, since the integral

$$\frac{1}{16\pi^2}\int_{R_3}\frac{e^{ik\{|x-z|+|z-y|\}}}{|x-z||z-y|}q(z)\,dz\equiv I(x,\ y,\ k^2)$$

converges absolutely and uniformly with respect to $x, y \in R_3$ when $\operatorname{Im} k \geq 0$, while the integrand is an integral function of k. In view of (3.8), the following estimate holds for $I(x, y, k^2)$ with $\operatorname{Im} k \geq 0$:

$$|I(x,\ y,\ k^2)|\leqslant C_0\alpha,$$

where

$$\alpha=\max_{x,\,y}\int_{R_3}\frac{dz}{(1+|z|^{3+\varepsilon})|x-z||y-z|}. \tag{3.10}$$

The integrand in the general term of series (3.6) is therefore a regular function of k in the half-plane $\operatorname{Im} k > 0$. Using Hadamard's relation and estimate (3.10), we obtain for $\operatorname{Im} k \geq 0$,

$$\left|I_{k^2}\begin{pmatrix}\xi_1\xi_2\ldots\xi_n\\ \xi_1\xi_2\ldots\xi_n\end{pmatrix}\right|\leqslant C_0^{2n+1}\alpha^n n^{\frac{n}{2}}\prod_{i=1}^{n}\frac{1}{1+|\xi_i|^{3+\varepsilon}}.$$

It follows from this that the integral appearing in the general term of (3.6) converges absolutely and uniformly and that the following estimate holds for the general term:

$$\left|\frac{(-1)^n}{n!}\int_{R_3}\int_{R_3}\ldots\int_{R_3}I_{k^2}\begin{pmatrix}\xi_1\xi_2\ldots\xi_n\\ \xi_1\xi_2\ldots\xi_n\end{pmatrix}d\xi_1\,d\xi_2\ldots d\xi_n\right|\leqslant\frac{n^{\frac{n}{2}}}{n!}C_0^{2n+1}\alpha^n\left[\int_{R_3}\frac{d\xi}{1+|\xi|^{3+\varepsilon}}\right]^n. \tag{3.11}$$

In view of what has been said above, the general term of (3.6) is a regular function of k for $\operatorname{Im} k > 0$. It can be seen from (3.11) that series (3.6) converges absolutely and uniformly and, consequently, its sum is a regular function of k in the half-plane $\operatorname{Im} k > 0$ and bounded right up to the real axis

$$\sup_{\operatorname{Im} k\geqslant 0}|D(k)|\leqslant\sum_{n=0}^{\infty}\frac{n^{\frac{n}{2}}}{n!}C_0^{2n+1}\alpha^n\left[\int_{R_3}\frac{d\xi}{1+|\xi|^{3+\varepsilon}}\right]^n<\infty.$$

In order to complete the proof of the lemma, it is necessary to show that series (3.6) can be differentiated term by term. It is sufficient to show that the series

$$\delta_{s,0}+\sum_{n=1}^{\infty}\frac{(-1)^n}{n!}\sum_{\substack{s_1+s_2+\ldots+s_n=s\\ s_i\geqslant 0}}\frac{s!}{s_1!s_2!\ldots s_n!}\times$$

$$\times\int_{R_3}\int_{R_3}\ldots\int_{R_3}\begin{vmatrix}I^{(s_1)}(\xi_1,\xi_1,k^2)\,I^{(s_1)}(\xi_1,\xi_2,k^2)\ldots I^{(s_1)}(\xi_1,\xi_n,k^2)\\ I^{(s_2)}(\xi_2,\xi_1,k^2)\,I^{(s_2)}(\xi_2,\xi_2,k^2)\ldots I^{(s_2)}(\xi_2,\xi_n,k^2)\\ I^{(s_n)}(\xi_n,\xi_1,k^2)\,I^{(s_n)}(\xi_n,\xi_2,k^2)\ldots I^{(s_n)}(\xi_n,\xi_n,k^2)\end{vmatrix}\prod_{i=1}^{n}q(\xi_i)d\xi_i, \tag{3.12}$$

obtained by an s-fold ($s \leq m$) term-by-term differentiation of series (3.6) converges absolutely and uniformly. We will use Hadamard's relation for the estimation of the expressions

$$\begin{vmatrix} I^{(s_i)}(\xi_1, \xi_1, k^2)\dots \\ \cdots\cdots\cdots \\ I^{(s_n)}(\xi_n, \xi_1, k^2)\dots \end{vmatrix} \prod_{i=1}^{n} q(\xi_i) \equiv \det_n \left| I^{(s_i)}(\xi_i, \xi_j, k^2) \right| \prod_i^n q(\xi_i), \tag{3.13}$$

so that we first estimate the quantities $I^{(s_i)}(\xi_i, \xi_j, k^2)$. In view of (3.8), we have for $\operatorname{Im} k \geq 0$

$$\left| I^{(s_i)}(\xi_i, \xi_j, k^2) \right| \leqslant \frac{1}{16\pi^2} \int_{R_3} \frac{(|\xi_i - \zeta| + |\xi_j - \zeta|)^{s_i}}{|\xi_i - \zeta||\xi_j - \zeta|} |q(\zeta)| d\zeta \leqslant$$

$$\leqslant \frac{6^{s_i - 1}}{16\pi^2} \int_{R_3} \frac{|\xi_i|^{s_i} + |\xi_j|^{s_i} + |\zeta|^{s_i}}{|\xi_i - \zeta||\xi_j - \zeta|} |q(\zeta)| d\zeta.$$

Without loss of generality we can consider that $C_0 \geq 1$. Then, using (3.10), we obtain the required estimate for $I^{(s_i)}(\xi_i, \xi_j, k^2)$, $\operatorname{Im} k \geq 0$, on account of (3.8)

$$\left| I^{(s_i)}(\xi_i, \xi_j, k^2) \right| < \frac{6^{2s_i - 1}}{16\pi^2} \left\{ |\xi_i|^{s_i} \alpha C_0 + |\xi_j|^{s_i} \alpha C_0 + \alpha C_{s_i} \right\}.$$

After a multiplication of both sides of the inequality by $\sqrt{|q(\xi_i)|}\sqrt{|q(\xi_j)|}$ the estimate takes the form

$$\left| \sqrt{q(\xi_i)} \sqrt{q(\xi_j)} I^{(s_i)}(\xi_i, \xi_j, k^2) \right| \leqslant$$

$$\leqslant \frac{6^{2s_i - 1}}{16\pi^2} \Big\{ \alpha C_0 \sqrt{|q(\xi_i)||\xi_i|^{2s_i}(1 + |\xi_i|^{3+\epsilon})} \sqrt{|q(\xi_j)|(1 + |\xi_j|^{3+\epsilon})} +$$

$$+ \alpha C_0 \sqrt{|q(\xi_j)||\xi_j|^{2s_i}(1 + |\xi_j|^{3+\epsilon})} \sqrt{|q(\xi_i)|(1 + |\xi_i|^{3+\epsilon})} +$$

$$+ \alpha C_{s_i} \sqrt{|q(\xi_i)|(1 + |\xi_i|^{3+\epsilon})} \sqrt{|q(\xi_j)|(1 + |\xi_j|)^{3+\epsilon}} \Big\} \times$$

$$\times \frac{1}{\sqrt{1 + |\xi_i|^{3+\epsilon}} \sqrt{1 + |\xi_j|^{3+\epsilon}}} \leqslant \frac{6^{2s_i} \alpha C_0^{\frac{3}{2}}}{16\pi^2} \left\{ \sqrt{C_{2s_i}} + C_{s_i} \right\} \times \frac{1}{\sqrt{1 + |\xi_i|^{3+\epsilon}} \sqrt{1 + |\xi_j|^{3+\epsilon}}}.$$

Now, using Hadamard's relation, we obtain the following estimate for the integrand for $\operatorname{Im} k \geq 0$:

$$\left| \det_n \left| I^{(s_i)}(\xi_i, \xi_j, k^2) \right| \prod_i^n q(\xi_i) \right| \leqslant \prod_i^n \frac{1}{1 + |\xi_i|^{3+\epsilon}} \times \frac{\prod_i 6^{2s_i}}{(16\pi^2)^s} \alpha^n C_0^{\frac{3n}{2}} n^{\frac{n}{2}} \times \prod_i \left\{ \sqrt{C_{2s_i}} + C_{s_i} \right\}.$$

This leads to the absolute and uniform convergence of the integrals appearing in the expression for the general term of series (3.12), as well as to the estimate

$$\left| \sum_{\substack{s_1 + s_2 + \dots + s_n = s \\ s_i > 0}} \frac{s!}{s_1! s_2! \dots s_n!} \int_{R_3}\int_{R_3} \cdots \int_{R_3} \det \left| I^{(s_i)}(\xi_i, \xi_j, k^2) \right| \prod_i^n q(\xi_i) d\xi_i \right| \equiv$$

$$\equiv a_n(k) \leqslant \left[\sum_{\substack{s_1 + s_2 + \dots + s_n = s \\ s_i > 0}} \frac{s!}{s_1! s_2! \dots s_n!} \prod_i \left(\sqrt{C_{2s_i}} + C_{s_i} \right) \right] \times \tag{3.14}$$

$$\times \frac{6^{2s}\alpha^n C_0^{\frac{3n}{2}}}{(16\pi^2)^n}\left\{\int\limits_{R_3}\frac{d\xi}{1+|\xi|^{3+\varepsilon}}\right\}^n \leqslant \left(\frac{\alpha\beta C_0^{\frac{3}{2}}}{16\pi^2}\right)^n 6^{2s} s! n^{\frac{n}{2}} \max_{\substack{\sum_{i=1}^{n} s_i=s\\ s_i>0}} \left(\prod_{i=1}^{n}\frac{\sqrt{C_{2si}}+C_{si}}{s_i!}\right)\sigma_s^n, \tag{3.14}$$

where $\beta=\int\limits_{R_3}\frac{d\xi}{1+|\xi|^{3+\varepsilon}}$ and σ_s^n is the number of ways we can select n positive numbers to sum to s. As can be easily shown, the number σ_s^n does not exceed $e^{n+s}F$, where F is a constant. Taking into account that

$$\sigma_s^0=\begin{cases}0 & \text{where } s>0,\\ 1 & \text{where } s=0,\end{cases}$$

and introducing the abbreviations

$$Q_{s,n}=\max_{\substack{s_i>0\\ s_1+s_2+\ldots+s_n=s}}\prod_i\frac{\sqrt{C_{2s_i}}+C_{s_i}}{s_i!},\quad A=\frac{\alpha\beta C_0^{\frac{3}{2}}}{16\pi^2},$$

we can rewrite estimate (3.14) as

$$|a_n(k)|\leqslant\left[A^n n^{\frac{n}{2}}e^n\right][Q_{s,n}][F6^{2s}s!e^s].$$

It should be noted that

$$Q_{s,n}\leqslant\max_{\substack{s_i>1\\ \sum s_i=s}}\prod_i\frac{\sqrt{C_{2s_i}}+C_{s_i}}{s_i}(C_0+\sqrt{C_0})^n\equiv Q(s)(C_0+\sqrt{C_0})^n.$$

It can be seen from this that series (3.12) is majorized by the convergent number series

$$\sum_{n=0}^{\infty}\frac{[Ae(C_0+\sqrt{C_0})]^n}{n!}n^{\frac{n}{2}}F(36e)^s s!Q(s),$$

so that series (3.6) can be differentiated term by term s times, $s\leq m$. In addition, we have obtained an estimate for arbitrary functions D(k), $\operatorname{Im} k\geq 0$,

$$|D^{(s)}(k)|\leqslant F_0 6^{2s}e^s s!Q(s),$$

where

$$F_0=F\sum_{n=0}^{\infty}\frac{[Ae(C_0+\sqrt{C_0})]^n}{n!}n^{\frac{n}{2}}.$$

The lemma has been proved.

Using the result obtained, we can easily extend some of the propositions of § 1 to the three-dimensional case.

Let us use the term "singular point of the operator L" for the complex or real positive number λ which is the root of the function $D(\sqrt{\lambda})$. The multiplicity of the singular point will be the multiplicity of the root. A point λ lying on the upper (lower) edge of the branch cut $[0, \infty)$ will be considered to be a singular point if $D(\sqrt{\lambda}) \to 0$ as $z \to \lambda$ along any path lying in the upper (lower) half-plane that is not a tangent to the real axis.

Let the set of all singular points of the operator L be denoted by E, the set of all nonzero eigenvalues by E_0, the set of all singular points lying on the edges of the cut $[0, \infty]$ by E_1, the set of all singular points of infinite multiplicity by E_2, the set of all points of accumulation of eigenvalues by E_3. Let $\tilde{E}$, $\tilde{E}_0$, etc., denote the images of these sets in the plane $k = \sqrt{\lambda}$.

In view of the well-known theorem of Kato [21], $E_0 \subset E \setminus E_1$. If the potential q(x) decreases faster than any power of $1/|x|$, then, according to Lemma 3.1, the function D(k) is regular in the upper half-plane $\operatorname{Im} k > 0$ and is continuous right up to the real axis together with all its derivatives. It follows from this that $E_3 \subset E_2 \subset E_1$. It should be noted further that the rank of an eigenvalue λ of the operator L (equal to the multiplicity of the corresponding pole of the resolvent) does not exceed the multiplicity of the corresponding singular point. This can be seen from formula (3.4).

The results concerning the spectrum of the operator L that can be obtained with the help of Lemma 3.1 can be formulated as two theorems.

Theorem 3.1

Let the potential q(x) of operator L be a continuous complex function satisfying the conditions

$$\sup_x |q(x)|(1+|x|^{3+\varepsilon})|x| < \infty,$$
$$\sup_x |\nabla q(x)|(1+|x|^{3+\varepsilon}) < \infty.$$

Then:

1) the set of eigenvalues is bounded and, having been numbered with rank taken into account, satisfies the condition

$$\sum \operatorname{Im} \sqrt{\lambda_\nu} < \infty;$$

2) $E_3 \subset E_1$;

3) the set E_1 is bounded, closed, is of zero measure, and satisfies the condition

$$\sum \ln |l_\nu| \, |l_\nu| > -\infty,$$

where $|l_\nu|$ is the length of the interval l_ν of contiguity with the set E_1 and the summation extends over all bounded intervals l_ν.

The proof of Theorem 3.1 is exactly the same as that of Theorem 1.1. Only the boundedness of the set E has to be demonstrated separately. In order to do this, we make use of the estimate given by Faddeev (see [22])

$$|I(x, y, k^2)| \leqslant C\frac{e^{-\operatorname{Im} k|x-y|}}{|k|} \leqslant \frac{C}{|k|} \text{ where } \operatorname{Im} k \geqslant 0.$$

Here C is a constant which depends only on sup |q(x)| and sup |∇q(x)|. Using the above estimate, we can replace (3.11) by

$$\left|\frac{(-1)^n}{n!}\int\limits_{R_1}\int\limits_{R_2}\cdots\int\limits_{R_n} I_{k^2}\begin{pmatrix}\xi_1, & \xi_2 \ldots \xi_n\\ \xi_1, & \xi_2 \ldots \xi_n\end{pmatrix} d\xi_1 d\xi_2 \ldots d\xi_n\right| \leqslant \frac{n^{\frac{n}{2}}}{n!} C_0^{2n+1}\alpha^n \left[\int \frac{d\xi}{1+|\xi|^{3+\varepsilon}}\right]^n \frac{1}{|k|^n}.$$

It can be seen from this that $D(k) - 1 = 0(1/k)$ and, consequently, the set of singular points of operator L is bounded.

Theorem 3.2

Let the potential q(x) of the operator L satisfy the conditions of Theorem 3.1 and, moreover, let

$$\sup_x |q(x)|(1+|x|^{3+\xi})|x|^s \leqslant C_s < \infty,$$

whatever the natural number s. Then, the set $\tilde{E}_2$ is bounded, closed, is of measure zero, and $\tilde{E}_3 \subset \tilde{E}_2$. Moreover, the following condition is satisfied:

$$\int\limits_0 \ln T(\sigma)\, d\varphi\left(\tilde{E}_2, \sigma\right) > -\infty, \tag{3.15}$$

where $T(\sigma) = \inf\limits_s F_0 6^{2S} e^S Q(s)\sigma^S$, F_0 and $Q(s)$ being the same as in Theorem 3.1, while $\varphi(\tilde{E}_2, \sigma)$ is the σ-vicinity measure of the set $\tilde{E}_2$.

The proof can be obtained directly if we make use of Lemmas 1.2 and 3.1.

Corollary 1. If the potential satisfies the condition

$$|q(x)| \leqslant C\exp(-\delta|x|^\alpha),\ \delta > 0,\ 0 < \alpha < \frac{1}{2},$$

then set $\tilde{E}_2$ is bounded, closed, is of measure zero, and satisfies the condition

$$\sum |l_\nu|^{\frac{1-2\alpha}{1-\alpha}} < \infty.$$

Here $|l_\nu|$ is the length of the interval l_ν of contiguity to the set $\tilde{E}_2$ and the summation extends over all bounded contiguity intervals.

Indeed, it is not difficult to establish that in the given case the estimate

$$\sup_x |q(x)|(1+|x|^{3+\varepsilon})|x|^s \leqslant Bd^s s^{\frac{s}{\alpha}},\quad s = 0, 1, 2\ldots$$

is valid; x, B, d are constants depending only on C, δ, α. Then, from Stirling's formula, we obtain

$$Q(s) = \sup_{\substack{\sum s_i = s\\ s_i > 1}} \prod_i \frac{1}{s_i!}\left(\sqrt{B}\, 2^{\frac{s_i}{\alpha}} + B\right) d^{s_i} s_i^{\frac{s_i}{\alpha}} \leqslant B_0 d_0^s s^{\frac{s}{\alpha}-s},$$

$$B_0 = B_0(C, \delta, \alpha), \ d_0 = d_0(C, \delta, \alpha)$$

and, consequently, we have

$$|D^{(s)}(k)| \leqslant B_1 d_1^s s^{\frac{s}{\alpha}}, \quad B_1 = B_1(C, \delta, \alpha), \ d_1 = d_1(C, \delta, \alpha). \tag{3.16}$$

After this the proof is exactly the same as that for the corresponding proposition of § 1 (see Corollary 1 to Theorem 1.2).

Corollary 2. If

$$\int\limits_{\delta} \ln T(\sigma)\, d\sigma = -\infty, \tag{3.17}$$

then the set $\widetilde{E}_2$ is a null set and, consequently, the total number of singular points of the operator L is finite and all singular points are of finite multiplicity. In particular, the number of eigenvalues of the operator L is finite. Condition (3.17) is satisfied, for example, when for any $C > 0$, $\delta > 0$ we have

$$|q(x)| \leqslant C \exp(-\delta \sqrt{|x|}).$$

In this case, the following estimate holds for the derivatives of the function D(k):

$$|D^{(s)}(k)| \leqslant B_1 d_1^s s^{2s}, \quad B_1 = B_1(C, \delta), \ d_1 = d_1(C, \delta).$$

From this we immediately obtain

$$\ln T(\sigma) \leqslant -\frac{\text{const}}{\sigma},$$

and, consequently, (3.17) is satisfied.

Literature Cited

1. M.A. Naimark, Dokl.Akad.Nauk SSSR, 85:41-44 (1952).
2. M.A. Naimark, Dokl.Akad.Nauk SSSR, 89:213-216 (1953).
3. M.A. Naimark, Tr.Mosk.Matem.Obshch., 3:181-270 (1954).
4. M.Sh.Birman, Matem. sb., 55(2) (1961).
5. B.Ya.Levin, Dokl.Akad.Nauk SSSR, 106:187-190 (1956).
6. B.S. Pavlov, Dokl.Akad.Nauk SSSR, 146:1267-1270 (1962).
7. V. P. Khavin, Investigations in Modern Problems in the Theory of Functions of the Complex Variable. Fizmatgiz (1961), pp. 121-131.
8. I.M. Gel'fand, Usp.Matem.Nauk, 7(6) (1952).
9. R.M. Martirosyan, Izv.Akad.Nauk Arm.SSR, seriya matem., 10(1) (1957).
10. N. Levinson, Kgl.Danske Videnskab Selskab, Mat.-Fys.Medd., 25(9) (1949).
11. L.D. Faddeev, Usp.Matem.Nauk 14(4) (1959).
12. Z.S. Agranovich and V.A. Marchenko, The Inverse Problem in the Quantum Theory of Scattering. Fizmatgiz (1959).
13. H. Weyl, Mat. Annalen, 68:220-269 (1910).
14. I.I. Privalov, Boundary Properties of Analytic Functions. GITTL (1950).

15. A. Beurling, Acta Math., Vol. 72 (1940).
16. G.E. Shilov, Mathematical Analysis. Fizmatgiz (1960).
17. P.I. Bushell, Quart.J.Math., 13:99-107 (1962).
18. V.A. Marchenko, Matem. sb., 52(2) (1960).
19. G.Polya and G.Szego, Problems and Theorems in Analysis, Vol. II, No. 4, Chapter 7 [Russian translation] GITTL (1956).
20. A.Ya. Povzner, Matem. sb., 32(1) (1953).
21. T. Kato, Comm. Pure Appl.Math, 12:403-425 (1959).
22. L.D. Faddeev, Vestn. Leningr.Gos.Univ., seriya matem., No. 7(2) (1957).